Cybersecurity

Cybersecurity

Current Writings on Threats and Protection

Edited by
JOAQUIN JAY GONZALEZ III *and*
ROGER L. KEMP

McFarland & Company, Inc., Publishers
Jefferson, North Carolina

Also of Interest and from McFarland

Legal Marijuana: Perspectiveson Public Benefits, Risks and Policy Approaches, edited by Joaquin Jay Gonzalez III *and* Mickey P. McGee (2019)

Eminent Domain and Economic Growth: Perspectives on Benefits, Harms and New Trends, edited by Joaquin Jay Gonzalez III, Roger L. Kemp *and* Jonathan Rosenthal (2018)

Small Town Economic Development: Reports on Growth Strategies in Practice, edited by Joaquin Jay Gonzalez III, Roger L. Kemp *and* Jonathan Rosenthal (2017)

Privatization in Practice: Reports on Trends, Cases and Debates in Public Service by Business and Nonprofits, edited by Joaquin Jay Gonzalez III *and* Roger L. Kemp (2016)

Immigration and America's Cities: A Handbook on Evolving Services, edited by Joaquin Jay Gonzalez III *and* Roger L. Kemp (2016)

Corruption and American Cities: Essays and Case Studies in Ethical Accountability, edited by Joaquin Jay Gonzalez III *and* Roger L. Kemp (2016)

Library of Congress Cataloguing-in-Publication Data

Names: Gonzalez, Joaquin Jay, III, editor. | Kemp, Roger L., editor.
Title: Cybersecurity : current writings on threats and protection / edited by Joaquin Jay Gonzalez III and Roger L. Kemp.
Other titles: Cybersecurity (McFarland & Company)
Description: Jefferson, North Carolina : McFarland & Company, Inc., Publishers, 2019 | Includes bibliographical references and index.
Identifiers: LCCN 2018058319 | ISBN 9781476674407 (softcover : acid free paper) ♾
Subjects: LCSH: Computer security. | Computer networks—Security measures. | Computer crimes—Economic aspects.
Classification: LCC QA76.9.A25 C925 2019 | DDC 005.8—dc23
LC record available at https://lccn.loc.gov/2018058319

British Library cataloguing data are available

ISBN (print) 978-1-4766-7440-7
ISBN (ebook) 978-1-4766-3541-5

Printed in the United States of America

McFarland & Company, Inc., Publishers
Box 611, Jefferson, North Carolina 28640
www.mcfarlandpub.com

Jay dedicates this book to Elise and Coral,
the next generation

Roger dedicates this book to Anika,
the best and the brightest

Acknowledgments

We are grateful for the support of the Mayor George Christopher Professorship at Golden Gate University, and GGU's Pi Alpha Alpha Chapter. We appreciate the encouragement from Dean Gordon Swartz and our wonderful colleagues at the GGU Edward S. Ageno School of Business, the Department of Public Administration, and the Executive MPA Program.

Our heartfelt "Thanks!" go to the contributors listed in the back section and the individuals, organizations, and publishers below for granting permission to reprint the material in this volume and the research assistance. We are particularly grateful to the International City/County Management Association or ICMA (icma.org), for permission to reprint a significant number of selections in this volume. The ICMA advances professional local government management through leadership, management, innovation, and ethics.

American Libraries
American Society for Public Administration
Anne Ford
Arun Vishwanath
Beth Payne
Benjamin Dean
Brian McLaughlin
Caitlin Cowart
Chelsea Binns
The Conversation
Dorothy Denning
e.Republic, Inc.
Federal Bureau of Investigation
Frank J. Cilluffo
Fred H. Cate
GCN Magazine
Golden Gate University Library
Governing
Government Finance Officers Association
Government Finance Review
Government Technology
James H. Hamlyn-Harris
John O'Brien
Kaiser Health News
Lichao Zhang
Mark Banks
Mark Kennedy
Michelle F. Hong
Mick McGee
NACo County News

National Association of Counties
Nir Kshetri
PA Times
Paul D. Harney
The Pew Charitable Trusts
PM Magazine
Public Management (PM) Magazine
Richard Forno
San Francisco State University
Scott Shackelford
Sharon L. Cardash
Texas Workforce Commission
Theodore J. Kury
U.S. Department of Commerce
U.S. Department of Homeland Security
U.S. Department of Justice
University of San Francisco Library
William Hatcher

Table of Contents

Part II. Threats and Risks

Part III. Prevention, Protection, Partnership

A. Practical Preventive Measures

III. *B. Technical Protections*

Preface

We live in a cyber age. The Internet of Everything (IoE) is here and growing. Big data is getting bigger and bigger. Billions of people are connected by billions of devices across cities and countries. While walking or driving, while in our homes or offices, wherever and whenever, personal and professional information are being received and transmitted. Selling to millions of customers and providing mass public services are now easily done through phones, tablets, and laptops. However, this widespread access has come at a significant sacrifice. Hackings and cybercrime have emerged as threats. Our businesses and governments are under siege and so are our privacy and democracy. How could security be restored in this complex cyber world?

This timely volume was researched with answers to this important question in mind. We compiled best practices and experiences from the evolving and dynamic field of cybersecurity. Our contributors, who come from diverse professions and backgrounds, report on the current and emerging threats and the risks as well as the available preventive measures and protections. These practical, easy-to-read cybersecurity lessons and suggestions are for individuals, organizations, and communities.

For ease of reading, and reference purposes, we divided this book into four major sections, which are highlighted below to provide the reader with detailed information about the content of this cybersecurity reference book.

Part I—An Overview

We begin this volume with a general overview of the digital dilemma we live and work in. Our contributors and we believe this rapidly evolving context is the source of both the good—and the bad and thus the need for cybersecurity.

The Internet of Everything, which encompasses people, devices, processes, and data, has created a lot of benefits but there are also a lot of costs. Chapters 1 through 5 describe the bad side of our electronic world. As giant repositories of citizen's information, governments, especially at the city and county level, are constantly under attack, sometimes without employees knowing it. The negative impact, especially cybercrime, has been increasing as well as the hacking risks.

But surely, it seems, the benefits far outweigh these risks and costs. After all, according to Chapters 6 to 9, moving to digital transactions and operations have made public and private services more responsive, open, speedy, and transparent. Moreover, advanced performance and data analytics have enhanced our capacity to process a cyber world where big data is now characterized with unprecedented volume, velocity, value, veracity, and variety. These daily data exchanges, e.g., texts, photos, tweets, are in the trillions. What are the privacy protections promised by public agencies and private companies? Are they really adequate?

Part II—Threats and Risks

Part II fleshes out the specifics of the "bad"—the many threats and risks that directly relate to individual, organizational, societal, and global cybersecurity. More than a dozen chapters (from 10 to 22) are included in this section of our national reference volume examining the threats of a hard to manage Internet of Things (IoT) industry and the risks of stockpiling Zero-Day software vulnerabilities.

We agree with our contributors who underscore both the quantity and quality of these threats and risks. Several chapters delve into the basics and experiences of database breaches, domestic and international hacking, covert espionage, denial of service attacks, phishing, spyware, ransomware, and the weak defenses of social media. Going even deeper, Chapter 15 distinguishes between the different types of hackers and hacks. Part II also includes graphic examples of cybersecurity lapses from court, healthcare, credit reporting, utility, transportation, and county organizations.

Part III—Prevention, Protection, Partnership

Part III is the core of this cybersecurity volume: 3 Ps—Prevention, Protection, Partnership. Nearly 30 colorful chapters with innovative preventive, and protective, and partnering lessons are presented. Readers are encouraged to copy those they feel would lead to appropriate corrective actions—making them, their organizations, and communities more secure and safe. Use them to reduce risks and combat threats.

Section A is a compilation of 11 practical preventive measures chapters (23 to 33) which sets forth a solid plan for cybersecurity, how to fight negative cybersecurity issues, steps to provide for online safety and security measures, and how to stay safe on social networking computer sites. Other chapters include a review of the steps that should be taken to enhance your cybersecurity practices, how to form a computer response team, and examples of cybersecurity threats that must be examined, dealt with, and properly responded to in a timely manner.

Section B deals with technical protections, starting with Chapter 34 on the country's current priority and response through the "National Cybersecurity Workforce Framework" which guides your organization towards building a world-class cyber workforce. Then continues with chapters (35 to 41) on relevant cybersecurity practices such as eliminating network "blind sports," suggested computer user responsibilities, and how to develop sound cybersecurity practices for users of an individual's and organization's computer hardware and software systems. Disaster recovery cybersecurity best practices are also surveyed.

Section C, the third and final portion of Part III, are chapters (42 to 50) which examine the use of creative cybersecurity partnerships and networks to enhance the safe usage of an individual's and organization's computer hardware and software systems and their applications. The partnerships and networks involve not only federal agencies but private sector stakeholders, from the state to cities and communities.

The cybersecurity initiatives of two U.S. presidents are also set forth and examined. They are President Obama's Cybersecurity Initiative and President Donald Trump's Executive Order on Cybersecurity. Other cybersecurity measures that are being used by our federal government, our nation's highest level of government, are also examined. The cybersecurity safety measures being developed and used by other levels of government, and their agencies, are also examined.

Part IV—The Future

The fourth, and final section of this volume, delves into the future of the dynamic and evolving field of cybersecurity. The half-dozen chapters (from 51 to 56) in this portion of the volume include bold predictions on what's coming up with cybersecurity, and report on Artificial Intelligence (AI) cyberattacks, 5G deployment talks, strategic thinking in the cyber domain, and far reaching internet, email, and computer use policies, as well as the problems with freezing your credit report.

The Appendices

Our Appendices are a repository of knowledge and lessons too. We developed a comprehensive glossary of cybersecurity terms to help you understand the technical jargon and practices. The reference section of this best practices volume also includes several outstanding cybersecurity public documents that could be mirrored by citizens and policymakers. These appendices include a glossary of cybersecurity terms, a copy of The Presidential Executive Order on Strengthening the Cybersecurity of the Federal Government, that is, the computer networks and the critical infrastructure that are related to it. A copy of the City and County of San Francisco, California, Cybersecurity Policy is also included in this section. The La Porte County, Indiana, IT Computer Security Policy is also appended for review and reference purposes.

We hope that the rich and colorful cybersecurity overview, assessments of threats and risks, 3P measures and actions, and predictions of future trends and practices contained in this volume will help inform and educate citizens, employees, businesses, advocates, and public officials, and facilitate their interest in the development of cutting edge cyber age devices and apps that are more secure. This will enhance the level of services they provide to everyone—individuals, citizens in general, as well as the public-at-large. The use of such cybersecurity practices will also hold down the cost of providing services, both public and private ones, to all citizens throughout the nation—both taxpayers that pay for government services, as well as citizens that pay for private services.

1. Government Employees Unaware They Are Cyber Crime Victims*

CHELSEA BINNS

Cyber crime is now the top threat facing America. Government personnel are key targets and are increasingly victimized. Actors are using stealth tactics to steal information without an employee's knowledge. However, many cyber breaches targeting government workers are preventable. It is important for government employees to understand how their information is stolen in order to inform preventive efforts. Recent cases of employee data breaches will be discussed to illustrate the threat.

According to recent reports, cyber crime costs the world economy between $375 billion and $575 billion annually and it has now replaced nuclear war as the most discussed global threat.

Government employees are especially vulnerable to cyber crimes. Government information security incidents have increased 1,121 percent in the last decade. Why? Many data thieves, including individual "hackers," large organized crime groups and nation states are especially interested in obtaining government information.

Their ability to obtain this information surreptitiously is growing. Sensitive information is stolen routinely from government employees, without their knowledge. This act, cyber espionage or cyber spying, is believed to have affected more than 800 million people during 2013. Experts believe rapidly expanding technologies are to blame. According to James Clapper, the

*Originally published as Chelsea Binns, "Government Employees Unaware They Are Cyber Crime Victims," *PA Times*, December 15, 2015. Reprinted with permission of the publisher and author.

director of National Intelligence, "The world is applying digital technologies faster than our ability to understand the security implications and mitigate potential risks."

In several cases, breaches are traced back to irresponsible employee behavior. In these instances, employees were unaware they perpetuated cybercrimes.

Per the Associated Press, in 2013, "About 21 percent of all federal breaches were traced to government workers who violated policies; 16 percent who lost devices or had them stolen; 12 percent who improperly handled sensitive information printed from computers; at least 8 percent who ran or installed malicious software; and 6 percent who were enticed to share private information."

One way government employees make themselves vulnerable to data breaches is by sending their government emails to a personal email account. Data thieves can easily obtain this information by hacking into the email account through a variety of methods, including social engineering. Ironically, employees who engage in this volatile practice risk breaching their own personal information.

This was seen in the case of CIA Director John Brennan, whose personal email account was hacked this year by a high school student. In this case, the perpetrator obtained classified agency information, which was maintained by Brennan in a personal email account. It was later learned that Brennan had "forwarded" classified documents to his private AOL email address, in violation of agency policy. One such document obtained was his 47-page application for top-secret security clearance. It has been reported the teen used social engineering methods to gain access into the email account.

Government employees also increase their vulnerability to data theft by clicking on unknown email links. Perpetrators design emails to appear legitimate, hoping employees will click on them. When they do, their data can be stolen without their knowledge. The email can release malware that steals the data. False links in emails can also bring employees to "fake" websites that steal their data. These are referred to as "phishing" emails.

A phishing email attack is believed to have caused a major United States Postal Service (USPS) data breach. In November 2014, *The Wall Street Journal* reported the records of more than 800,000 people, including "employees, top directors and regulators," were exposed in a USPS hacking that was traced back to China. It was later determined that employee medical records were among the sensitive information exposed in the breach.

It was learned employees contributed to the hacking by clicking on phishing emails. Following the breach, the Inspector General (IG) for the USPS conducted a test to determine how employee behavior contributed to the hacking. The USPS IG sent emails containing false links to 3,125 Postal

Service employees. Most employees, 93 percent, did not report the emails, per policy. In addition, 25 percent of employees clicked on the fraudulent email and most of those employees, 90 percent, did not report that fact to the IG.

These cases illustrate the opportunity for government employees to prevent agency data breaches. Employees are urged to exercise extreme caution when handling agency emails.

To this end, government employees should also take their agency's training courses to learn how to best handle agency data. Employee training may have prevented the breaches at the USPS. Following their test, the USPS IG learned that most employees who "failed" had not completed the required information security training.

As illustrated, the cyber threat to government employees is serious and becoming increasingly problematic. Recent cases have demonstrated the ability of data thieves to steal government information from employees who are unaware of the risk and/or have not exercised caution when handling agency data. It is clear that employees can reduce the risk through taking simple precautions.

2. Cybersecurity and America's Governments*

Roger L. Kemp

Cyber security deals with computer security and the protection of both an organization's computer hardware and software systems. "Cybersecurity" describes a dynamic and evolving effort to protect an organization's hardware and software from things called viruses, bugs, worms, eavesdropping, spoofing, phishing, clickjacking, social engineering, etc. As this field has evolved, and computer systems and their software have become more sophisticated, these virus vehicles have been combined to create the new common title of cybersecurity. This term also deals with the protection of the information on an organization's computer system. Computer hackers can steal sensitive and confidential information, such as names and addresses, credit card information and medical information, stored on an organization's computers.

Computers have evolved in recent years. While every organization has a central mainframe computer, every work station now has a desktop computer, and each employee typically has a laptop computer, as well as a handheld computer (smart phone or tablet). Computer hardware, over the years, has become smaller, more sophisticated, less expensive and more user-friendly. Software, on the other hand, has become available in more fields, is more sophisticated, more user-friendly and the training of employees to operate computer applications (or apps) provided by newly acquired software is now common practice in all government organizations, regardless of their size.

Cyber risk is a major threat to all government organizations, including cities and counties, states, as well as our federal government. There is a high risk in many organizations that do not have the resources to have an infor-

*Originally published as Roger L. Kemp, "Cybersecurity and America's Governments," *PA Times*, August 15, 2017. Reprinted with permission of the publisher and author.

mation technology department, or the like. The process of recouping losses after a computer hack is burdensome and costly, with a lot of resulting litigation, so cyber risk needs to be properly addressed in all public organizations, for their employees, as well as the public that they serve, to protect from such expenses.

Increasingly, data breaches and cyber extortion practices are taking place daily, and protecting public computers and their data is of primary importance. Local, state and federal government organizations need to take control and focus on such important matters as:

- Understanding what you have on your computer systems that need protection;
- Encrypting data and devices, which is the first line of defense on any cybersecurity plan;
- Establishing and implementing the best cybersecurity practices by putting in place cybersecurity protocols and procedures for all employees in the organization to follow; and
- Obtaining and periodically reviewing cyber-insurance policies to be sure your organization is protected by having adequate insurance coverage.

Whatever new hardware and software is acquired by a government organization, its managers need to be sure they are educating all of their employees on the specifics and providing regular scheduled system updates. The majority of data breaches happen by accident or mistake, with the employees being one of the greatest causes of such breaches. Most attacks are phishing attacks and take place when employees click on links or attachments. It is essential to educate everyone in your organization on their computer hardware and software, and to enact appropriate protocols to add an extra layer of protection for the computer hardware and software systems used in your organization.

The constant threat of a cyber attack is the most important problem for our local, state and federal governments, who generally do not know how often they are attacked, or what kinds of attacks are taking place on their organization's computer hardware and software systems. Studies suggests that, on the average, local governments in the United States are not doing the kind of job necessary to achieve high levels of cybersecurity on their organization's computer hardware and software systems.

Previously, data processing was primarily done in the Finance Department and the Department of Public Works, which usually has an Engineering Division. Over time, more and more departments got computers and programs. Eventually, some larger governments formed computer related management departments. The names of these new departments have evolved, and some government managers called the services they provide to user departments by different names. The title of this department has evolved in recent years, and some of the more common titles of this evolving department are highlighted below for the reader's information:

- Information and Communications Technology (ICT) Department,
- Information Technology (IT) Department,
- Information Resources Management (IRS) Department,
- Information Systems (IS) Department,
- Enterprise Technology (ET) Department,
- Management Information Services (MIS) Department, and
- No doubt that other departmental titles will evolve in future years.

The term cybersecurity is dynamic and evolving, and its implementation, or lack thereof, impacts all levels of government organizations. Our nation's professional membership organizations should be congratulated for their state-of-art efforts to bring forth the latest best practices in the dynamic and evolving field of cybersecurity for government public officials throughout our nation. These best practices will benefit their organizations, its employees, as well as the public they serve.

3. NACo Members Zero In on Cybersecurity, Tech Safety*

Ariel Cohen

Cybersecurity has become a topic of national and international consequence in recent months, but local governments are also affected by online hackers. County leaders recently discussed readiness to respond to such attacks and how to update their systems at the 2017 NACo Legislative Conference.

Ninety percent of all data that exists today has been generated since 2010, and most of that data is not properly secured, said conference speaker Bill Wright, director of government affairs and senior counsel at Symantec. The public sector is especially at risk for cybercrime, as government sites are ranked fourth among reported cyber-attacks in the past year, he said.

"You cannot expect perfection. As a county official, you have to say 'When we get hit' not 'If we get hit,'" said Steve Hurst, director of safety and security strategies at AT&T. "Figure out the steps you will take so you can go immediately into a data-disaster recovery plan the second there is a breach," he added.

Hurst, along with Rita Reynolds, CIO of the County Commissioners Association of Pennsylvania, helped county leaders create a plan to work through likely cyber threats. Reynolds emphasized that no matter your perceived level of risk, every county should have a plan ready to go. If your local government is unsure how to create a plan, most states have templates for a county to follow.

---

*Originally published as Ariel Cohen, "NACo Members Zero In on Cybersecurity, Tech Safety," *NACo County News*, March 20, 2017. Reprinted with permission of the publisher.

- Before doing anything else, county leaders should determine which department has the most valuable data and prioritize its protection, the experts advised. While every department will argue its data is most important, determine what is most necessary to protect the safety of your citizens.
- Hiring a third party to perform a risk assessment of your county's tech systems and data can help you do the difficult work of prioritizing needs, Hurst and Reynolds advised.
- Local officials should also keep an eye on the growing dominance of crypto-ransomware. Symantec's Thomas MacLellan encouraged counties to pay closer attention to cybersecurity on devices that are a part of the "Internet of Things," such as high-tech refrigerators, toasters or even self-driving cars. "I look at cybersecurity as infrastructure at this point," MacLellan said. "It's no different than a road, it's no different than a school, and it's going to cost money. We're going to have to get over that."

"Ransomware has evolved from an annoyance to a serious threat," Wright added. "If you see something that says 'Click here to watch Justin Bieber get punched in the face' you can't do it. Tempting as it is, even for a guy like me, you just can't click it."

While counties can work to shield themselves from hacking, local government can't completely shield their constituents from attacks. In his speech to NACo members during the opening general session, Virginia Sen. Mark Warner, vice chairman of the Senate Intelligence Committee, warned county officials to be wary of the media influence Russian hackers could have on their districts.

"If you searched for "hacking in the election," you wouldn't get Fox News or NBC or CNN; you would literally get—for the first five stories—Russian propaganda.... You'd get stories about Hillary Clinton being sick," Warner said. "If we don't get our arms around it, it's only going to get worse."

The rapid spread of information can be very useful, but disastrous when used incorrectly.

Jake Williams, manager of strategic initiatives at StateScoop, explained that it has become too easy to spread misinformation or "fake news," on social media, and counties would be wise to monitor information circulating online about their communities.

Many counties don't focus enough on their social media presence or do not have a large enough budget to monitor all social media channels. Of the 85 percent of local governments that use social media platforms to connect with their constituents, 55 percent do not track or monitor their social media interactions, according to Public Technology Institute. Not only does monitoring social media channels help stop the spread of "fake news," but it also helps county officials keep a pulse on the needs of their residents.

In the past year, many communities have grappled with the growing shared economy and the technological and regulatory challenges that come with it. Oftentimes, these "shared economy" services, such as Airbnb, Uber and EatWith draw in new users with a simple registration process, which is often far easier than registering a new business with the government, said Tim Woodberry, director of government affairs at Accela.

Woodberry said that "things like EatWith straddle the line between friends getting together and an unlicensed restaurant." While each county handles the issue differently, he suggested that simplifying the government licensing process would improve local compliance rates.

4. Growing Impact of Cybercrime in Local Government*

GERALD CLIFF

A direct result of the ever-increasing reliance on technology and the Internet is that governments at all levels are rapidly becoming dependent on these technologies for provision of essential services. The downside to this reliance is that it magnifies the risk of cyber intrusions and data breaches.

These breaches can result in the compromise of personally identifiable information (PII) of every resident whose name, date of birth, and Social Security number reside on a local government server. Governmental entities also maintain records of personal health information (PHI) pertaining to their employees that, if compromised, could expose employees to serious problems.

Data breaches from phishing, hacking, and insider threat are on the increase and causing considerable damage in terms of costs to seal the breach and address the potential damage to those whose PII has been compromised. Here are examples:

In April 2015, the Florida Department of Children and Family Services (DCF) suffered a data breach when a state employee used the employee's employment-related access to obtain the personal information of thousands of Floridians.

According to the Department of Economic Opportunity (DEO), one of its employees managed to access the Florida Department of Children and

---

*Originally published as Gerald Cliff, "Growing Impact of Cybercrime in Local Government," *Public Management (PM) Magazine*, June 2017 issue. Reprinted with permission of the publisher.

Families' Florida ACCESS system. He then obtained the names and Social Security numbers of more than 200,000 people in the DCF system. In March 2015, the DEO employee was arrested and charged with alleged trafficking and unauthorized use of PII.

In May 2016, more than 100 Los Angeles County employees fell prey to a phishing scam, revealing usernames and passwords that were then used to disclose personal information of approximately 756,000 individuals who had done business with county departments.

The *2016 IBM Cost of Data Breach Report* finds the average consolidated total cost of a data breach grew from $3.8 million to $4 million. The study also reports that the average cost incurred for each lost or stolen record containing sensitive and confidential information increased from $154 to $158 per record.

Ransomware on the Rise

Crypto-ransomware attacks software that encrypts a victim's data and then offers to sell the victim the decryption key are on the rise and have already crippled hospitals, police departments, educational systems, critical municipal infrastructure, and other vital cornerstones of the public sector. Examples include:

- An April 2016 ransomware attack against the Lansing Michigan Board of Water and Light crippled the agency's ability to communicate internally and with its customers, and ultimately cost the city-owned utility about $2 million for technical support and equipment to upgrade its security.
- In June 2016, a police department in Collinsville, Alabama, refused to pay the ransom and lost access to a database of mugshots.
- The Cockrell Hill, Texas, police department was attacked in December 2016. The department thought that it could restore files; however, the files were not properly backed up and the department lost eight years' worth of digital evidence, including some documents, spreadsheets, video from body-worn and in-car cameras, photos, and surveillance video.
- In February 2017, the Roxana, Illinois, Police Department fell victim to a ransomware attack. Although the department refused to pay the ransom, the incident still cost the city considerable time and money. Rather than pay the ransom, the department was forced to "wipe the system." The department had backups of all important information, though restoring those backups in a usable manner is requiring significant manpower.

In a 2016 whitepaper on the topic of ransomware, the Osterman Research Corporation stated "both phishing and crypto ransomware are increasing at the rate of several hundred percent per quarter, a trend that it is believed will continue for at least the next 18 to 24 months." The report went on to point out the "FBI estimates that ransomware alone cost organizations $209 million in just the first three months of 2016."

According to former U.S. Federal Bureau of Investigation Director James Comey, "Cybercrime is becoming everything in crime. Again, because people have connected their entire lives to the Internet, that's where those who want to steal money or hurt kids or defraud go."

Various foreign entities, such as China, Russia, and North Korea, are known to be engaged in hacking activities. But there are also literally millions of individuals who operate on their own by purchasing hacking tools on the dark Web.

According to the *2015 Trustwave Global Security Report*, attackers receive an estimated 1,425 percent return on investment ($84,100 net revenue for each $5,900 investment in software and tools). Cybercrime is becoming easier and safer to commit thanks to the relative anonymity of the Internet and the availability of hacking tools for purchase, also referred to as "malware as a service" schemes.

This means that literally, anyone with a computer and Internet access can become a hacker, as programming and networking skills are no longer a requirement.

Legal Liability

The immediate damages caused by a breach are just the tip of the iceberg. A class-action lawsuit from residents whose credit card information was exposed through a local government's online fee payment system will hurt, but relatively few of them will have suffered direct, unreimbursed losses. Their losses were generally absorbed by their banks and insurance companies.

What do you tell the voters who's PII and PHI is exposed to identity thieves through the actions (or inaction) of their state or local government? How do you deal with the loss of public trust that accompanies this type of event?

As the holder of that confidential information, there comes a level of responsibility to take proper precautions to protect it. Breach that responsibility and there may be consequences.

Think of how the information your organization collects and processes could be used to commit a crime:

- PII could be used to commit identity theft.
- Personal information on children, witnesses, informants, victims of crimes, and other vulnerable populations could be used to violate their privacy or facilitate crimes against them.
- Information on regulated business could be used by business rivals.
- Information on government bidding, contracting, or economic development plans could be used to competitive advantage.
- Information dealing with active investigations (civil, criminal, or administrative) could be used to compromise those investigations.

Personal information on government employees could be used to exact revenge for unpopular decisions or actions.

Insurance Loopholes

A governmental insurance policy can provide protection; however, in the face of the extreme costs of a breach, insurance companies may find ways to decline coverage.

If the insurance policy doesn't specifically provide coverage for data breach-related damage, or if the insured agency fails to adhere to the details specified by the insurer pertaining to security measures, the insurer is likely to refuse payment.

Insurance companies may agree to cover a governmental entity for losses caused by cyberintrusions and data breaches, but there will be requirements that must be met by the insured. Where a claim is submitted, an investigation will ensue, and where security requirements have not been met, a claim will likely be denied.

In what appears to be a landmark case, Cottage Health System of California suffered a data breach in which 30,000 records containing PII were exposed in a cyber incident. A class action suit in 2014, filed on behalf of the affected clients, resulted in a $4.125 million award against the health system, in state court.

Although Cottage Health had a cyber insurance policy with Columbia Casualty, the insurer pointed out that the insured stored the confidential information on an Internet accessible system, but failed to install encryption or other safeguards. The insurer denied the claim. Imagine if your locality were suddenly confronted with that kind of financial loss.

To effectively mitigate the dangers involved in data breach and cyber intrusion, it is important to understand as much as possible about these types of attacks. In an effort to better understand the impact these crimes are having on state, local, territorial, and tribal government, the National White Collar Crime Center (NW3C) has for several years assembled a data set of incidents using publicly available data from a wide range of sources.

Analyzing the Data

As of January 1, 2017, NW3C's data set of state and local governmental entities reporting data breaches totaled slightly more than 1,900 incidents. The organization's analysis of this information points to some important differences between the public and private sectors.

Recall that a data breach does not require a cyberintrusion; it can involve the exposure of confidential information through lost paper files, improperly disposed of electronic devices with digital memory, even intentional theft.

A cyberintrusion also may not necessarily involve a data breach. Technically hacktivism is where a hacker illegally accesses a computer network for the purpose of defacing or otherwise interfering with that network is a cyberintrusion but does not involve the breach of confidential PII and PHI.

When examined across all sectors, the data breaches resulting from cyberintrusions or hacking within the past 10 years account for less than 30 percent of the more than 5,000 data breaches reported by the Privacy Rights Clearinghouse. This includes not only the public sector but also retail, education, and health sectors.

Analysis of the state/local-specific data that NW3C compiled showed the percentage of government cyber incidents involving hacking differs somewhat from that of the combined sectors of retail, education, healthcare, and nonprofits. What stands out most in our analysis is that insiders in government-related data breaches played a far greater role.

Analysis of the cases in NW3C's dataset indicates the government sector shows a 48.05 percent greater propensity (difference between 17.4 percent and 33.5 percent) to suffer from unintentional insider breaches and a 48.55 percent greater propensity to suffer from malicious insider breaches, while the probability of suffering a data breach through lost or stolen devices is 64.53 percent less in the government sector.

The insider seems to play a noticeably greater role in data breaches in government than in the other sectors commonly tracked. Regardless of why public employees appear more prone to causing data breaches, the importance of our findings is that it facilitates identification of potential solutions that can potentially help reduce the incidents of data breach.

Reducing Risk

Due to space limitations and the extensive list of remedial actions available that could mitigate incidents of data breach and cyberintrusion, a comprehensive program is far too much to address in the space allotted for this article. We can, however, outline several basic steps, which can serve as the starting point for a comprehensive program that could reasonably be expected to result in reducing an agency's exposure to risk.

There is an array of technical tools that an IT coordinator can choose from to help ensure network security; however, in addition to those tools, a manager should consider the personnel-related aspects of security. Here are some basic steps.

1. Establish detailed and thorough policies pertaining to Internet use. Encrypt e-mails and other content containing sensitive or confidential data. Enforce rules regarding access to personal social media accounts in the course of the workday. Direct the IT coordinator to be responsible for the monitoring of all communications for malware. Control the use of personally owned devices that are able to access corporate resources.

2. Implement best practices for user behavior. Employees must select passwords that match the sensitivity and risk associated with their data assets. Employee passwords not only must meet certain criteria pertaining to strength, but also must be changed on a regular basis. IT departments should be required to keep software and operating systems up-to-date to minimize malware problems. Employees should receive thorough training about phishing and other security risks, and they should be tested periodically to determine if their anti-phishing training has been effective. Employees whose duties involve off-site Internet access, should be trained in best practices when connecting remotely, including the dangers of public Wi-Fi hotspots.

3. Maintain a timely and complete backup of your critical systems.

4. Regularly practice restoring your system from those backups. It is important to be aware that there are strains of ransomware that can be programmed to activate on a time delay, so backups may end up including the latent ransomware program. A careful manager needs to be aware that backups alone may not be effective unless they have been thoroughly checked and determined to be safe.

5. The Intelligence National Security Alliance (INSA 2013) practice recommends that a risk-reduction program includes an insider threat component that encompasses, at a minimum:

- Organization-wide participation.
- Oversight of program compliance and effectiveness.
- Confidential reporting mechanisms and procedures to report insider events.
- An insider threat incident-response plan.
- Communication of insider threat events.
- Protection of employees' civil liberties and rights.
- Policies, procedures, and practices that support the insider threat program.
- Data collection and analysis techniques and practices.
- Insider threat training and awareness.
- Prevention, detection, and response infrastructure.
- Insider threat practices related to trusted business partners.
- Inside threat integration with enterprise risk management.

For more information on the question of legal liability and what you can do to limit your exposure to the threat of cybercrime, see the NW3C white paper *Cyberintrusions and Data Breaches* that can be found at http://www.nw3c.org/research.

5. Cybersecurity*

What's Your Risk?

International City/County Management Association

"Cybersecurity is an iceberg topic: The largest part is what you don't see—and that's the part that can sink an organization," wrote Wayne Sommer, internal audit manager in Aurora, Colorado, and former director of administration and finance at ICMA. Sommer was writing for the ICMA Blog, and his suggestions were so good that we're repeating them for people who didn't see the blog post.

One of Sommer's responsibilities in Aurora is to look at his organization through the eyes of "risk," the possibility of an event or condition occurring that will have an impact on the ability of the organization to achieve its strategic objectives. With that in mind, he teamed up with Tim McCain, Aurora's information security officer, to come up with a list of questions managers should ask as they address the risk of cyber disruptions.

"In Aurora, we are watching the cybersecurity ice mountain grow larger and larger before our eyes, but we are not sitting still or ignoring it," McCain wrote. "We are looking below the surface at the complex issues involved and chipping away at them strategically, consistently, and in line with the resources we have at hand. It's not an easy task. It's a big berg. We have no choice; neither do you."

Cybersecurity affects *everyone.* Large? Medium? Small? Regardless of your organization's size, you are a potential target. E-mail scams, network attacks, and ransomware are just three of the predators out there looking for

*Originally published as International City/County Management Association, "Cybersecurity: What's Your Risk?," ICMA Website, August 9, 2016. Reprinted with permission of the publisher.

a vulnerable target. Cities take advantage of technology to make internal operations and service delivery more efficient and effective. The "Internet of things," which enables cities to use the Internet in ways never before imagined, provides an exponential number of increasing opportunities for mayhem.

A manager can hand the problem off to IT and move on, but the problems and the solutions go beyond IT. Cybersecurity is an *organizational* issue that just happens to enter in through the technology door. Addressing the issue needs to start at the top and involve everyone within the organization.

Managers deal with risk every day, whether they are conscious of it or not. And cybersecurity is a risk challenge that requires a conscious approach. Here are six questions a manager can use as a framework to get started.

1. ***What could go wrong?*** Brainstorm. No possibilities are off the table. The more voices in the room from every staff position and generation, the more likely you are to gather a comprehensive list of possibilities.

2. ***What would be the early warning signs?*** How would you know if something is amiss? Have your staff noticed more mysterious e-mails in their inboxes? Is your IT Department finding anomalies showing up on their reports? Is your administrative staff receiving any odd phone calls? Identify as many warning signs as you can and try to understand if you have the ability to monitor them and alert the appropriate people if they occur.

3. ***What is the likelihood of this event or condition occurring?*** This is somewhat subjective, but have your staff consider your existing defenses, the status of your hardware and software, and even your computer use policies. Did you know that any staff member sitting in front of a monitor and keyboard is your greatest vulnerability point? Open one wrong e-mail attachment and … well, it could get ugly quickly. Gauging your threat awareness and readiness will help you estimate the likelihood of an event occurring. No one is immune. The question is no longer "if" you will ever get hacked, but "when."

4. ***What would be the impact if an event did occur?*** First thoughts here jump to financial hits and that is a real possibility; but don't forget to consider potential impacts on internal operations, external service delivery, and especially reputation impacts. Who wants to work for an organization that cannot keep its employees' personal information out of the public domain?

5. ***How would you respond if it did occur?*** This is critical. Once an event occurs, how you respond can affect the severity of the impact. Your response can also boost or further destroy your organization's reputation in the public's eye. Identify the response resources required—

time, money, and people—for which you should plan before an incident occurs.

6. ***What are you doing now that would minimize the impact or likelihood of this risk or condition if it did occur?*** You can use your work in determining your organization's current preparedness level to address this question.

Sommer and McCain suggest gathering a cross-section of your operational staff to begin answering these questions related to cybersecurity. That's a good starting point that can help you evaluate your preparedness, identify your worst vulnerabilities, and provide a basis for generating an action plan to begin addressing this critical issue. "Start now," Sommer says. "That iceberg is headed your way."

Contacts: Wayne C. Sommer, internal audit manager, Aurora, Colorado (wsommer@auroragov.org); Tim McCain, information security officer, Aurora, Colorado (tmccain@auroragov.org).

6. Performance and Data Analytics*

The Cincinnati Story

Harry Black

I have been driven my entire career to show that government can operate at a high performance level. Government organizations vary in terms of their level of performance and overall effectiveness, with some local governments optimizing performance better than others. Why is that? This case study was prepared to provide insight to that question.

Government is inherently not designed to be efficient, swift nor nimble. There are several built in challenges that must be overcome to truly optimize performance. Civil service systems are outdated, and most local governments must also manage the challenges of labor contracts. Additionally, government organizations are limited in terms of performance incentives that can be offered to the workforce.

These challenges are realities. However, the challenge is to engage local government management in such a way that these issues can be overcome by integrating sound labor management principles and practices. With performance and data analytics, this can happen.

Time is of the essence for local governments in this age of the internet of things and big data. A fundamental question is "Can government be disrupted?" The answer is "yes" and it will happen. Driverless cars, drones, big data driven algorithms and robots will serve as the source of that disruption. The way that we currently conduct business in govern-

---

*Originally published as Harry Black, "Optimizing Local Government Management Through Performance and Data Analytics: The Cincinnati Story," *Government Finance Review*, June 2017. Reprinted with permission of the publisher.

ment will change. It will require fewer people, fewer facilities and less equipment.

Disruption can be minimized and/or averted, but only if government aggressively pursues an examination of what it does and how it does it. It must daily ask itself several basic questions, "Are we making a difference? "Are we maintaining and exceeding our customer's expectations? "Are we optimizing innovation? "Are the impactful things that we are doing sustainable?"

Optimizing local government management through performance and data analytics can strengthen local government, while also enhancing its relevancy.

The Cincinnati Experience

In 2014, my phone rang while I was loading groceries into my car. It was an executive recruiter calling to talk about the city manager opening in Cincinnati, Ohio. At the time I was somewhat hesitant to pursue the opportunity. I was satisfied with what I was doing—serving as the Chief Financial Officer for the City of Baltimore. However, since becoming a city manager had always been a career goal I decided to pursue the opportunity. Saying "yes" led to the most grueling, yet fulfilling recruitment process I ever experienced.

I made three trips to Cincinnati and met with over 150 business, neighborhood and religious leaders. The discussions were illuminating and various themes began to resonate with me. All those I met with shared a desire for an effective, efficient and responsive city government that would:

- Improve customer service;
- Be more responsive;
- Improve economic inclusion;
- Overcome infrastructure challenges;
- Reinvent the City's permitting process; and
- Enhance safety of the city.

It became clear Cincinnati was ripe for performance and data analytics. During my interview sessions with the Mayor and City Council Members, I began to broach the subject and was pleased that they all were quite receptive. I knew that if I was selected I would aggressively pursue instituting a comprehensive performance and data analytics program. This is something that I attempted in the past while serving in senior management roles for other cities, but generated limited success. In those cities my role, although significant, was not as significant as the role of City Manager.

When the executive recruiter called to tell me that I had been selected, I would have the unique opportunity to show Cincinnati could buck the prevailing views and perceptions of government.

Government does not have to be characterized by terms such as mediocre, slow and inadequate customer service. Local government management can be innovative, collaborative, interactive, transparent and high performing. We can make a huge difference in the lives of children and families while also growing the local economy.

I believe that performance and data analytics can quickly assist local governments with optimizing overall performance, generating numerous economies and efficiencies, as well as operational breakthroughs.

Moving forward, this paper will focus on the City's experience using performance and data analytics to make Cincinnati government more nimble, strategic, responsive and transparent. In short, performance and data analytics has allowed Cincinnati to make a difference in the lives of its citizens. I will also share some of our successes to date.

Cincinnati has a local government supporting a residential population of 300,000. It does this with 6,400 employees spanning 25+ departments and offices, and a $1.4 billion budget. Cincinnati provides a full range of municipal services including police and fire protection, parks and recreation, highways/streets, waste collection, health and human services, culture, planning and zoning, and water/sewer services.

What we have endeavored to achieve in Cincinnati is a comprehensive integrated approach systematically integrating several critical components, including:

- One Page Strategic Plan
- Performance Management Agreements
- CincyStat
- Performance Budgeting
- Innovation Lab
- Open Data

I took the helm as City Manager in September 2014 and my first order of business was to prepare a business case to share with the Mayor and City Council, requesting funding to establish what is now the City's Office of Performance and Data Analytics. City Council approved funds for the creation of the office in October 2014. The positions of Chief Performance Officer and Chief Data Officer were established, and between October 2014 and May 2015 we built a stand alone, state-of-the-art facility which became the Office of Performance and Data Analytics (OPDA).

The program officially launched in May 2015. By housing the core of our program in OPDA–CincyStat, Innovation Lab and Open Data—we are able to maximize collaboration. This is a unique approach, in that other cities rarely connect these functions. Our goal in having pursued these initiatives has been to make Cincinnati the best managed city in America by using our resources better, faster and smarter.

OPDA has a cumulative impact that pays for itself over time as a result of:

- Improved quality of customer service and reduced turn-around times
- Direct cost reductions/avoidance
- Revenue enhancements
- Goodwill

OPDA's role is five-fold:

- Facilitate transparency and accountability
- Understand City operations
- Creatively and strategically problem solve
- Optimize performance
- Find opportunities for improvement
- Nurture enterprise-wide collaboration

The program has achieved several milestones since its inception. OPDA was established to develop and lead performance initiatives. The One Page Strategic Plan has been adopted and supports the setting of administrative priorities. We have designed, developed and deployed Performance Management agreements with each department head to set priorities and expectations. An Innovation Lab has been built and operationalized for process streamlining (think LEAN and Six Sigma). We are hosting bi-weekly CincyStat performance management sessions. We are using Open Data for posting municipal datasets to the public in traditional formats, as well as in visual dashboard formats. We have even dabbled in predictive analytics to apply data science tools to increase effectiveness.

One Page Strategic Plan

Although not well known, Cincinnati serves as the home of multiple Fortune 500 and Fortune 1,000 corporate headquarters. This means we have a plethora of talent and intellectual resources at the ready. One of the many people I met during this recruitment process was a retired Procter & Gamble executive who now works with organizations across the country to develop strategic plans. He put me and my leadership team through a one page strategic plan process, using the One Page Solutions OGSP tool. The One Page

Solutions strategy emphasizes clarity and purpose and helped us hone in on what success looked like. And more importantly, it helped avoid putting together a typical three-ring binder plan that collects dust on a shelf. The One Page Strategic Plan has five sections:

1. Mission: Concise statement of why we are here or what we do.
2. Objective: What does success look like?
3. Goals: Metrics which will track progress versus the objective.
4. Strategies: The How
5. Plans: The most important projects/actions that define each strategy.

Through this process we were able to establish five priority goals:

1. Innovative Government
2. Fiscal Sustainability and Strategic Reinvestment
3. Thriving and Healthy Neighborhoods
4. Safe Streets
5. A Growing Economy

The Plan was one of my First 100 day goals and we were off to a good start. I have always been a big believer in how you start is typically how you finish. Your first 100 days generally determine how your first year will likely go, and your first year will generally determine how your overall tenure will go.

Performance Management Agreements

Something that I was eager to experiment with was the creation of performance management agreements entered into by the City Manager and individual department heads. This is a new concept for the most part in that I am only aware of one or two other municipalities that have pursued something similar.

These agreements are tied directly to our One Page Strategic Plan, integrated into the City's budget process, the employees annual review and published. Since the introduction of these agreements, we have established 100 department level priority initiatives. In addition, there are over 1,500 data points identified for regular collection and reporting by City departments, ensuring all ships always are sailing in the same direction.

Innovation Lab

Our Innovation Lab (I-Lab) is a collaborative facility to help redesign and streamline municipal processes to deliver better, faster, economical and smarter service. Our team will identify and scope projects prior to an I-Lab event. Facilitators help apply LEAN and Six Sigma principles to optimize efficiency and effectiveness. In the I-Lab everyone is equal and all viewpoints are sought out. The ILab experience can be rejuvenating and often evokes passion, which is good for team building and camaraderie.

We have had many successes as a result of the I-Lab, with one of the biggest being the streamlining of our building permit review and approvals process, reducing City approval times in half from 10.5 weeks to 3 to 5 weeks. We also conducted permit fee analysis to ensure fees are competitive, and target fee increases toward complicated projects to add resources devoted to streamlining interagency coordination.

Another success involved eliminating utility bill late fees. At one time the City received nearly 300 utility bills. However, we were consistently late with payment because there was no process in place to manage this, and we would incur about $133,000 annually in late fees. The I-Lab shed light on the process. We no longer pay late fees and have realized productivity gains as a result of not having 300 people involved in the paying of these bills.

CincyStat

CincyStat is our primary tool used to drive performance and strategic outcomes. It is a leadership strategy to mobilize City agencies to produce specific results. The Chief Performance Officer leads a series of regular, periodic meetings with the City Manager and leadership team, and each department's leadership. The meetings use data to analyze past performance, set new performance objectives and examine overall performance strategies.

There are Four Core Tenets that characterize Stat programs:

1. Accurate and timely intelligence shared by all.
2. Effective tactics and strategies.
3. Rapid deployment of resources.
4. Relentless follow-up

Tenet Four, in my view, is the most important.

A traditional Stat room consists of a podium for the agency head and agency staff to address questions from the panel. The panel consists of the City Manager, Assistant City Managers, Chief Performance Officer, and the heads of our Budget, Finance, GIS, Human Resources. Law and IT departments. There are two projectors that project charts and other information contained in a particular Stat memo for everyone to see. Our sources of data are various databases, our Customer Service Request system, and our Geographic Information system. Software applications aid us in organizing data, as well as assist us with visualizing it. Our performance analysts also conduct field work that is integrated into the Stat process.

For every meeting a comprehensive executive briefing memorandum is prepared, which serves as the focal point for discussion. It provides status updates on recurring operations, short-term and long-term projects. It also allows us to monitor core operations using key performance indicators. We delve into specific issues with background information, analysis, charts and questions in order to find opportunities for improvement.

Once a Stat session with a department concludes, our performance analyst prepares a follow-up memo to the department summarizing the session and identifies follow-up items to be addressed at the next Stat session.

Since OPDA's inception, Cincinnati's performance management programs have had a profound impact on improving service delivery and overall efficiency. The City has been able to eliminate Customer Service Requests backlogs related to our Transportation and Engineering and Public Services departments. We have been able to achieve an initial 7% increase in average overall customer satisfaction through the use of feedback from over 1,400 surveys completed.

Open Data

The City launched its new CincyInsights website in early 2017, which provides a showcase for a wide range of interactive public dashboards. These dashboards provide anyone with internet access an opportunity to review City data by way of user-friendly visualizations. These dashboards take existing City data already found in the City's Open Data Cincinnati portal and translates the content into graphical heat maps and charts. Users are able to interact with, and easily analyze mapped data using filters such as neighborhood location, date, activity type and more.

Currently the CincyInsights website features more than 15 dashboards that contain various datasets. Dashboards range from real-time snow plow tracking information to in-progress road projects to heroin overdoses. Each dashboard is organized according to our five strategic priorities. Additional visualizations will be added over time.

Giving this tool to the general public encourages individuals and groups to develop creative ways to engage, improve and serve the community. The CincyInsights project is an extension of the City of Cincinnati's overall commitment to transparency and data-driven government innovation.

Perhaps the single most powerful tool that makes this possible is the City's Geographic Information System (CAGIS). It is an enterprise-wide information system that provides access to real-time data for decision support, leading to improvements in the coordination, efficiency and quality of public service. The system embeds existing business rules and the management of information resources directly into departmental workflows, all made possible through the innovative integration of geographic information system (GIS) technology with automated business-process workflow software.

Cincinnati's data strategy, deployed city-wide, ensures transparency and enhanced customer service through frequent publication of high quality data for public consumption while enhancing performance management.

Conclusion

Through these initiatives founded in the principles of performance and data analytics the City has been able, and is positioned well to continue, to enhance customer service delivery, increase accountability and stimulate economic activity through information sharing.

This is only made possible through strong executive leadership starting with the Mayor and City Council who have embraced this approach from day one. Additionally, the thousands of employees who have contributed greatly in developing and implementing these changes are the real heroes and are to be commended.

OPDA has generated a 7-to-1 return on investment and has enhanced fiscal monitoring and financial oversight.

Just as Cincinnati has used these methods to optimize management effectiveness, so can other governments. It works. It is helping us meet and exceed the expectations of our residents and all those who live, work and play in Cincinnati.

7. The Digital Solution*

SHAUN MULHOLLAND

PM, October 13, 2015

In 2013, Allenstown, New Hampshire (4,300 population), began to digitize all of its paper files and streamline the town's administrative processes. Before this, paperwork and processes were antiquated and inefficient.

The town initially focused on its accounts payable process. It formed a LEAN team—the LEAN process was developed by Toyota Corporation to enhance efficiency by eliminating wasted time and processes—to study the current accounts payable processes and develop new ones where needed. In other words, more value for residents with fewer town resources used. It then addressed the digitization of public records, including meeting minutes, property files, and public access to agenda packets for the town's various boards.

The LEAN analysis indicated several limiting factors. The use of paper documents resulted in wasted time and resources, and the inability of staff and officials to access paper documents in a timely fashion added time to complete processes. Such costs as paper, copying, filing, mailing, check processing, and transporting documents were adding up for the town.

The most economical and efficient solution was a paperless system; however, this required digital processes that were new to the staff. Technological solutions needed to be added to Allenstown's growing IT infrastructure to achieve desired efficiencies.

The town implemented a virtual private cloud for all departments. It selected a vendor for this process, which allowed for all departments to work collaboratively on a common platform.

---

*Originally published as Shaun Mulholland, "The Digital Solution," *Public Management (PM) Magazine*, November 2015 issue. Reprinted with permission of the publisher.

Prior to this, departments operated on independent computer servers located in each of their respective buildings. These systems were eliminated when programs and data were transferred to the cloud.

Electronic Signatures

Another major factor that impaired efficiency was the need for signatures on various documents. The town solved the problem by implementing an electronic signature process, which relies on two providers.

RightSignature is the primary Web-based tool used to sign documents electronically. SeamlessGov is another solution used for creating documents with an electronic signature application.

The Electronic Signatures in Global and National Commerce Act (ESIGN) passed by Congress on June 30, 2000, allows for the use of electronic signatures in commerce and government activities.

There are, however, some key documents in which a party can require inked signatures, including contracts. A review of 15 U.S.C. 7001 (United States Code, 2006 Edition, Supplement 4, Title 15–COMMERCE AND TRADE) is advised when considering the implementation of electronic signature processes.

Digital Signatures

Most documents executed by Allenstown's officials are done by electronic signature. This allows for a document to be reviewed by multiple signatories at the same time from their respective locations.

Members of the board of selectmen are able to sign documents remotely with town-issued computer tablets. Documents originating in paper form, which are scanned as documents that are solely digital, are amenable to electronic signature processes.

Electronic signatures provide such security features as a checksum for detecting digital errors or a digital fingerprint. Arguably, these digital security measures are more effective than inked signatures. The checksum provides the date and time the document was signed and the IP address of the device on which the document was signed.

It's a common perception that an inked signature is easier to verify. The reality is quite different. Proving a forgery case based upon handwriting analysis is less than a perfect science. Handwriting experts also are hard to find. This is especially true in most states that rely upon state-level forensic laboratories for analysis.

There are states—as is the case in New Hampshire—that do not have a handwriting analyst at the state police forensic lab. They rely upon the FBI lab, which has threshold requirements and a long waiting period.

The second issue for accounts payable was the method by which the town received invoices and paid them. Invoices were being received by mail in a paper format. Checks would then be cut and mailed to the vendor.

To change this process, the town created a dedicated e-mail account to receive invoices from vendors, and it also implemented electronic fund transfers to pay them. With this method, we were able to receive invoices and pay vendors with a turnaround time of seven days or less. Prior to that, vendors were paid on average within 30 days.

We also worked with vendors to agree to send the town electronic invoices and transfer payment to their bank accounts by electronic fund transfer. Most vendors agreed to participate in this process, as it allowed for faster receipt of payment.

The accounts payable process we use now is documented by digital files only. The process took approximately seven months to implement. There were several phases of the process: identification of problem, analysis of solutions, testing of solutions, training of personnel, and actual implementation.

Despite the somewhat detailed and lengthy process, we still had problems that had to be addressed after implementation; fortunately, these were relatively minor. One of the issues was the criticality of naming conventions for the documents to allow for ease of search.

Another issue was the need for Adobe Acrobat Professional software, which allows for the ability to combine documents. An example would be a scanned invoice and a Word or scanned purchase order that need to be combined into one document. I would recommend having a detailed process in place before venturing into such a major transition.

Digitization of Property Files

While it streamlined the accounts payable process, the town contracted with Ricoh to digitize all property files. This included planning, zoning, and building files. Ricoh picked up the Allenstown files at town hall and transported them to their office facility.

It scanned the files into PDF documents, and then indexed and transferred them to our cloud. This step required minimal input from staff to address files and documents, which the vendor was unable to interpret.

All departments now have access to property files; before these files had only been available in paper format at town hall. This availability is particularly useful to firefighters, police officers, and public works personnel in the field, as they can access the files on a 24/7 basis.

Digitization of the files also allows instant access by staff at town hall as well as personnel in the field and at other town facilities. This access reduces staff time and the ability to have access to information more quickly, allowing for decisions to be made faster.

We frequently receive requests from the public for these documents. Previously, a staff person had to manually search through various paper files for requested documents. This took time and generally required a staff person copying the documents to provide them to the requesting party.

Quick Search

Staff can now search for the file by address, map and lot number, or key word search and quickly e-mail documents to the requester. The files are optical character reader (OCR) enabled, meaning that typed, handwritten, or printed text is converted into machine-encoded text allowing for the search of key words within PDF documents.

The speed with which the staff can search and provide those documents to a requestor saves the town money and allows the individuals faster access to needed information. The best thing about this is that it also enhances the speed by which commerce can be transacted.

The next phase of the public documents process is still under review. Allowing access to the files by the general public 24/7 through a Web portal on our website would be the next step, allowing the public to access documents when they wish to without the need for assistance from town staff.

We are currently examining two solutions. Documall and SeamlessGov have Web applications for this purpose. Documall has a solution that has been in use for some time, and SeamlessGov has a pilot program with Jersey City, New Jersey, currently under way.

We are still in the exploratory phase of our analysis and will need to carefully look at the cost-benefit delta—ratio of the change in price to the increase in value provided by the service—before we implement this next phase of our efficiency project.

The system Allenstown created is not a software or Web application. It is simply a set of folders listed under a property folder on one of our drives on the cloud. The folders are identified by map/lot number and address. Each of the 1,900 properties within the town has its own folder. The applicable documents for each property are located within the individual folders.

This is a relatively low-tech solution as well as low in cost to implement and maintain. Communities have software applications for document management. Ours is low cost and a relatively inexpensive solution, which may be attractive for smaller jurisdictions with limited budgets.

A similar process was used for accounts payable as the accounting software that was being used was unable to store the digital documents. The system of electronic folders we created was far more efficient than paper folders and achieved measurable cost savings.

The accounting system we had was unable to allow for electronic signatures in the approval process. We have since transitioned to Tyler Technologies' Infinite Visions software. It provides both electronic signature approval processes and storage of invoices, statements, and shipping documents within the accounting software application.

Public Meeting Documents

When I served as a department head, I remember attending various board meetings and listening to board members discuss community issues. They had access to the applicable documents that were being referenced; however, members of the public could only ascertain bits and pieces of the issue because they did not have access to the same documents. This was a noted shortfall in our effort to achieve more open government.

Our consolidated town website managed by Virtual Towns & Schools (VTS) provides a tool that allows for a system of indexed documents to be attached to the agenda for all board, committee, and commission meetings. This allows the staff to make these documents available to the public when the agenda is posted on the website.

VTS also has a subscriber option that allows members of the public to subscribe to agendas posted for various meetings at the website. When an agenda is posted, the subscriber receives an e-mail with a link to the agenda and attached indexed documents for the meeting.

Residents who attend board meetings with their tablets and smartphones are able to access these documents at the meetings. We also find this is a helpful tool for our land-use board meetings. Owners of property that abuts with another property, commercial interests, and residents are all able to review plans and proposals prior to and during the meeting.

Previously, someone would either have to come to town hall to review the plans or attend a meeting and crowd around the paper plans to get an understanding of what was being proposed. This provides another tool to allow for meaningful participation by informed residents, and it is another step in our overall goal of making more information available to the public faster.

Beyond Technology

I think most local government managers feel the pressure to try to keep up with the latest hardware and software when it comes to technology. We have applied an approach in Allenstown that is not specifically focused on the implementation of technology, but more on the analysis of efficiency.

Solutions to problems that can increase efficiency in many cases involve the implementation of new technologies—at least new to us in Allenstown. We have found through staff analysis of various issues that technology is usually just part of a solution.

The staff has also found in several cases that although a new technological solution may provide increased efficiency, it may not be the most cost effective to implement at a particular time.

8. Digital Wallpapers Open Doors*

CAITLIN COWART

Libraries are employing unique methods to make their digital collections available to patrons outside of the library. As a part of its Digital Library Community Project, San Antonio Public Library (SAPL) created digital wallpapers—virtual bookshelves that give patrons access to ebooks by simply scanning a QR code with a smartphone—that can be placed throughout the community. SAPL Community and Public Relations Manager Caitlin Cowart explains how the library developed the system.

SAPL launched the Digital Library Community Project in 2014 to spread awareness and create a gateway to SAPL's digital collection in physical spaces across our community.

The first project to launch was the digital library wallpaper, which is a two-dimensional adhesive resembling a bookshelf that can be temporarily applied to a wall without damage. From a marketing perspective, the wallpapers are interactive tools that serve as an introduction and gateway to the materials, while creating awareness about SAPL. Each book selection has a QR code that links to a title in our OverDrive collection as well as to titles in the public domain.

Perennially popular books are featured—*The Goldfinch* by Donna Tartt, *The Racketeer* by John Grisham, and *Gone Girl* by Gillian Flynn, for example. We also include books from local school reading lists, Spanish-language selections, and kids' books like *Diary of a Wimpy Kid*. An SAPL library card is required to access the OverDrive collection; non-cardholders can access only public domain titles.

---

*Originally published as Caitlin Cowart, "Digital Wallpapers Open Doors," *American Libraries*, March 1, 2017. Reprinted with permission of the publisher.

The wallpaper design was created by SAPL's graphic design team and was inspired by cellphone company Vodafone, which made ebooks available on its phones as a part of a publicity campaign in Romania to showcase the phone's capabilities. Once SAPL's original design was created, the graphic design team worked with the library's in-house digital services department to create QR codes that link to specific titles in our collection. We also worked with collection development to purchase more copies of the titles featured on the wallpapers to avoid or minimize holds. The final step was working with a local company to print and install the wallpapers, which cost around $900 each, on average. The San Antonio Public Library Foundation and the Friends of the San Antonio Public Library funded the project.

The initial launch of partners to display the wallpaper included a local YMCA, senior recreation and community centers, the Haven for Hope homeless shelter in downtown San Antonio, and our own Central Library. Twenty-three locations now feature the wallpapers, including the Henry B. González Convention Center, the DoSeum (a museum for kids), and numerous public park facilities.

The Digital Library Community Project is an effective and visually appealing way to market our services. Digital library usage has increased 50% since its launch. SAPL had more than 1 million checkouts through OverDrive in 2016. The project also serves as a discovery tool and gateway for those who are not yet using technology and apps. A part of our responsibility as a library is to help lead the way and teach people about technology. We have a digital divide in our community. According to U.S. Census and Pew Research data, 39% of San Antonio residents do not have broadband internet access at home, and 24% have no internet connection at all. The Digital Library Community Project is one of the many tools we have engaged to help solve that problem.

9. Personal Privacy Is Eroding as Consent Policies of Google and Facebook Evoke "Fantasy World"*

FRED H. CATE

We live in a world increasingly dominated by our personal data.

Some of those data we choose to reveal, for example, through social media, email and the billions—yes, billions—of messages, photos and Tweets we post every day.

Still other data are required to be collected by government programs that apply to travel, banking, and employment and other services provided by the private sector. All of these are subject to extensive government data collection and reporting requirements.

Many of our activities generate data that we are not even aware exist, much less that they are recorded. In 2013, the public carried 6.8 billion cell phones. They not only generate digital communications, photos and video recordings, but also constantly report the user's location to telephone service providers. Smartphone apps, too, often access location data and share them through the internet.

Added to the mix are video and audio surveillance, cookies and other technologies that observe online behavior, and RFID chips embedded in passports, clothing and other goods—a trove of data collected without our awareness.

---

*Originally published as Fred H. Cate, "Personal Privacy Is Eroding as Consent Policies of Google and Facebook Evoke 'Fantasy World,'" *The Conversation*, December 15, 2014. Reprinted with permission of the publisher.

Trillions of Transactions a Year

Much of this data is aggregated by third parties we've never heard of with whom we have limited or no direct dealings. According to *The New York Times*, one of these companies, known as Acxiom, alone engages in 50 trillion data transactions a year, almost none collected directly from individuals.

Known as information intermediaries, they calculate or infer information from demographic information such as income level, education, gender and sex; census forms; and past behavior, such as what clothes and foods someone purchased. That can generate data profiles that can be very revealing and used in determining credit scores, marketing predictions and other ways to quantify us.

As the volume, importance and, indeed, the value of personal data expand, so too does the urgency of protecting the information from harmful or inappropriate uses. But as we know, that's not easy.

Most data protection laws in the U.S. and elsewhere place some or all of the responsibility for protecting privacy on individual subjects through what's called "notice and consent."

In 1998, for example, the U.S. Federal Trade Commission, after reviewing the "fair information practice codes" of the United States, Canada and Europe, reported to Congress that the "most fundamental" principles to protect privacy are "notice" and "consumer choice or consent."

U.S. statutes and regulations tend to parallel the FTC's rules and recommendations on notice and choice. All U.S. financial institutions are required to send every customer a privacy notice every year, and doctors, hospitals and pharmacies provide similar notices, usually on every visit.

The focus on notice and consent is not limited to the United States. The draft of the European Union's General Data Protection cites "consent" more than 100 times and emphasizes its importance.

People across the world are spending more and more of their lives online, in the process handing over billions and billions of documents with their personal data.

All Our Fault

The truth is that notice and consent laws do little to protect privacy but typically just shift the responsibility for protecting privacy from the data user to the data subject—that would be us. After all, if anything goes wrong, it is our fault because we consented—often without realizing it.

Individual consent is rarely exercised as a meaningful choice. We are all overwhelmed with many long, complex privacy policies that most of us never read.

It is no wonder. One 2008 study calculated that reading the privacy policies of just the most popular websites would take an individual 244 hours—or more than 30 full working days—each year.

A reliance on notice and choice both under-protects privacy and can interfere with and raise the cost of beneficial uses of data, such as medical research and innovative products and services. (This is especially true when personal information is used by parties with no direct relationship to the individual, generated by sensors or inferred by third parties.)

"Fantasy World"

In a May 2014 report, the U.S. President's Council of Advisors on Science and Technology described the "framework of notice and consent" as "unworkable as a useful foundation for policy." The report stressed that "only in some fantasy world do users actually read these notices and understand their implications before clicking to indicate their consent."

There are better alternatives. One is enacting laws that place substantive limits on risky or harmful data uses, for instance. Another is to increase oversight by government and self-regulatory agencies, which could potentially forbid certain uses of personal data by third parties.

Many privacy advocates note that the U.S. is the only industrialized country without a dedicated privacy office in the federal government. Creating one might help ensure more attention is paid to privacy.

Other efforts are underway to restrict notice and choice to times when they are necessary and meaningful, and then to make them simpler and clearer.

Another promising approach would be to ensure that businesses take responsibility for their uses of personal data by making them legally liable for the reasonably foreseeable harm they cause, rather than allowing them to use notice and consent to continue shifting the responsibility to us.

At minimum, big users of personal data should be required to assess and document the risk those uses pose, and the steps they have taken to mitigate those risks. A more formal approach to managing privacy risks could better protect privacy, lead to greater consistency and predictability over time, and allow data users to make productive uses of data if risks can be mitigated.

The alternative is to continue to rely on notices no one reads, choices no one understands, and the other ineffective tools of the fantasy world that privacy law has become.

10. IoT Is Changing the Cybersecurity Industry*

LARRY KARISNY

It's odd that the Internet of Things (IoT) industry—an industry with a dismal record of cyberbreaches—would be the one moving cybersecurity forward, but that is exactly what is happening. With regulation looming and the bad press from recent breaches, there is no longer a choice: Better IoT security is a must. I will be speaking at the IoT Evolution Expo in Orlando this month on this very subject, and thought I'd give you a sneak peek.

IoT Security Gets a Failing Grade

If I remember correctly, 50 percent is a failing grade and yet, nearly 50 percent of IoT companies reported some type of security breach in recent memory. This shocking reality confirms that something needs to be done to improve IoT cyberdefense—and quickly. I cover a lot of areas in cybersecurity and know of no other industry with such a bad track record of breaches. Though these hacks expose data, not all of it has value. Sometimes an IoT hack garners useless data and offers no intelligence to use in an exploit, denial of service or machine control attack. The better news is that there are, at last, cyberdefenses coming to market that can address the need for solid IoT security.

*Originally published as Larry Karisny, "IoT Is Changing the Cybersecurity Industry," *Government Technology*, January 16, 2018. Reprinted with permission of the publisher.

New Cyberdefense Technologies Needed for IoT

IoT is different and has the potential to change everything. It is the new extended edge that allows unprecedented applications and intelligence with tremendous economics and accuracy. These tiny devices are the next step in physical artificial intelligence (AI). I stress "physical." They are out in the real world telling both people and machines what they need to know and need to do. If hacked, they can manipulate or destroy physical things with impacts that can extend to entire economies or worse cause loss of life. IoT is not just a database. IoT it is an actuary in the physical world that must be authenticated, validated and secured or risk the potential for very real danger.

Deep IoT Needs Deep Security

There's no room for a standard encryption file sizes or even simple processor updates patches in IoT. These tiny devices were built around minimal battery life that required tiny low-powered processors with minimal flash memory. This limitation has pushed the entire cybersecurity industry to rethink how we currently secure all digital technologies. We are beginning to see the successful deployment of these new security technologies today. If we are going to have deep learning in artificial intelligence and IoT we need to have deep security as well. IoT is pushing new security technologies toward achieving this goal.

The Enhanced Blockchain IoT Security Fit

Today's centralized security models require high infrastructure and maintenance cost associated with centralized clouds, large server farms and networking equipment. The sheer amount of communications that will have to be handled when IoT devices grow into the tens of billions will create bottlenecks and points of failure that can disrupt the entire network. Decentralized blockchain technologies could address these limitations, though blockchain alone is not a complete solution. As a principal in a company offering enhanced blockchain security, I am aware that blockchain alone is promising, but it is not the total answer. Just like current layered security architectures today, what we need in blockchain is a secure and safe IoT where privacy is protected. Enhanced blockchain-layered security technologies can offer this.

Revolutionize or Regulate

It is always better to self-regulate, and I hope the IoT industry gets that opportunity to find security solutions on its own. In working with cybersecurity entrepreneurs, I find that compliance and regulation seem to never catch up to the pace required by cyberdefense technologies. Billions were spent in security compliance of the smart grid. And while these security guidelines have value, at the end of the day, compliance does not mean you are secure. Hackers change things daily while compliance recommendations can take years. Cyberdefense needs to be more proactive, as does the matured working technologies that need to be used.

Preparing for Post-Quantum

Quantum computing and IoT have a very bright future. I stress "future" because there are a lot of issues that need to be addressed prior to quantum computing and IoT working together. Quantum computing in the short-term though will have the processing power to crack any static encryption algorithm. Solutions of more complex encryption algorithms with larger files sizes will work for IoT or really any other industry. In my last article, *Is Cybersecurity Encryption Ready to Break?*, I discussed the importance of looking for new low-overhead encryption technologies.

The IoT Security Opportunity

IoT suppliers that have a future will be the ones that invest in the security of their products. Even venture capital startups are clearly aware that they need to secure their IoT applications. If they do not, they could lose customers, spend money on regulatory issues or, worse yet, be involved in legal action against them. The smart IoT suppliers are embarrassing security and advertising it, even if it involves a premium price. They are beginning to find that customers will pay the premium. There are even IoT enterprise, managed services and cloud computing companies getting into the game offering their own solutions. IoT security is not a matter of choice anymore, it is a requirement.

11. Equifax Breach Is a Reminder of Society's Larger Cybersecurity Problems*

RICHARD FORNO

The Equifax data breach was yet another cybersecurity incident involving the theft of significant personal data from a large company. Moreover, it is another reminder that the modern world depends on critical systems, networks and data repositories that are not as secure as they should be. And it signals that these data breaches will continue until society as a whole (industry, government and individual users) is able to objectively assess and improve cybersecurity procedures.

Although this specific incident is still under investigation, the fact that breaches like this have been happening—and getting bigger—for more than a decade provides cybersecurity researchers another opportunity to examine why these events keep happening. Unfortunately, there is plenty of responsibility for everyone.

Several major problems need to be addressed before people can live in a truly secure society: For example, companies must find and hire the right people to actually solve the overall problems and think innovatively rather than just fixing the day-to-day issues. Companies must be made to get serious about cybersecurity—at a time when many firms have financial incentives not to, also. Until then, major breaches will keep happening and may get even worse.

---

*Originally published as Richard Forno, "Equifax Breach Is a Reminder of Society's Larger Cybersecurity Problems," *The Conversation*, September 21, 2017. Reprinted with permission of the publisher.

Finding the Right People

Data breaches are commonplace now, and have widespread effects. The Equifax breach affected more than 143 million people—far more than the 110 million victims in 2013 at Target, the 45 million TJX customers hit in 2007, and significantly more than the 20 million or so current and former government employees in the 2015 U.S. Office of Personnel Management incident. Yahoo's 2016 loss of user records, with a purported one billion victims, likely holds the dubious record for most victims in a single incident.

In part, cybersecurity incidents happen because of how companies—and governments—staff their cybersecurity operations. Often, they try to save money by outsourcing information technology management, including security. That means much of the insight and knowledge about how networks and computer systems work isn't held by people who work for the company itself. In some cases, outsourcing such services might save money in the short term but also create a lack of institutional knowledge about how the company functions in the long term.

Generally speaking, key cybersecurity functions should be assigned to in-house staff, not outside contractors—and who those people are also matters a lot. In my experience, corporate recruiters often focus on identifying candidates by examining their formal education and training along with prior related work experience—automated resume scanning makes that quite easy. However, cybersecurity involves both technical skills and a fair amount of creative thinking that's not easily found on resumes.

Moreover, the presence (or absence) of a specific college degree or industry certification alone is not necessarily the best indicator of who will be a talented cybersecurity professional. In the late 1990s, the best technical security expert on my team was fresh out of college with a degree in forest science—as a self-taught geek, he had not only the personal drive to constantly learn new things and network with others but also the necessary and often unconventional mindset needed to turn his cybersecurity hobby into a productive career. Without a doubt, there are many others like him also navigating successful careers in cybersecurity.

Certainly, people need technical skills to perform the basic functions of their jobs—such as promptly patching known vulnerabilities, changing default passwords on critical systems before starting to use them and regularly reviewing security procedures to ensure they're strong and up to date. Knowing not to direct panicked victims of your security incident to a fraudulent site is helpful, too.

But to be most effective over the long term, workers need to understand more than specific products, services and techniques. After all, people who understand the context of cybersecurity—like communicating with the public, managing people and processes, and modeling threats and risks—can come from well beyond the computing disciplines.

Being Ready for Action

Without the right people offering guidance to government officials, corporate leaders and the public, a problem I call "cyber-complacency" can arise. This remains a danger even though cybersecurity has been a major national and corporate concern since the Clinton administration of the 1990s.

One element of this problem is the so-called "cyber insurance" market. Companies can purchase insurance policies to cover the costs of response to, and recovery from, security incidents like data breaches. Equifax's policy, for example, is reportedly more than U.S.$100 million; Sony Pictures Entertainment had in place a $60 million policy to help cover expenses after its 2014 breach.

This sort of business arrangement—simply transferring the financial risk from one company to another—doesn't solve any underlying security problems. And since it leaves behind only the risk of some bad publicity, the company's sense of urgency about proactively fixing problems might be reduced. In addition, it doesn't address the harm to individual people—such as those whose entire financial histories Equifax stored—when security incidents happen.

Cybersecurity problems do not have to be just another risk people accept about using the internet. But these problems are not solved by another national plan or government program or public grumbling about following decades-old basic cybersecurity guidelines.

Rather, the technology industry must not cut corners when designing new products and administering systems: Effective security guidelines and practices—such as controlling access to shared resources and not making passwords impossible to change in our "internet of things" devices—must become fundamental parts of the product design process, too. And, cybersecurity professionals must use public venues and conferences to drive innovative thinking and action that can help fundamentally fix our persistent cybersecurity woes and not simply sell more products and services.

Making Vulnerability Unprofitable

Many companies, governments and regular people still don't follow basic cybersecurity practices that have been identified for decades. So it's not surprising to learn that in 2015, intelligence agencies were exploiting security weaknesses that had been predicted in the 1970s. Presumably, criminal groups and other online attackers were, too.

Therefore, it's understandable that commercialism will arise—as both an opportunity and a risk. At present, when cybersecurity problems happen, many companies start offering purported solutions: One industry colleague called this the computer equivalent of "ambulance chasing." For instance, less than 36 hours after the Equifax breach was made public, the company's competitors and other firms increased their advertising of security and identity protection services. But those companies may not be secure themselves.

There are definitely some products and services—like identity theft monitoring—that, when properly implemented, can help provide consumers with reassurance when problems occur. But when companies discover that they can make more money selling to customers whose security is violated rather than spending money to keep data safe, they realize that it's profitable to remain vulnerable.

With credit-reporting companies like Equifax, the problem is even more amplified. Consumers didn't ask for their data to be vacuumed up, but they are faced with bearing the consequences and the costs now that the data have gotten loose. (And remember, the company has that insurance policy to limit its costs.)

Government regulators have an important role to play here. Companies like Equifax often lobby lawmakers to reduce or eliminate requirements for data security and other protections, seek to be exempted from liability from potential lawsuits if they minimally comply with the rules and may even try to trick consumers into giving up their rights to sue. Proper oversight would protect customers from these corporate harms.

Making a Commitment

I've argued in the past that companies and government organizations that hold critical or sensitive information should be willing to spend money and staff time to ensure the security and integrity of their data and systems. If they fail, they are really the ones to blame for the incident—not the attackers.

A National Institute of Standards and Technology researcher exemplified this principle when he recently spoke up to admit that the complex password requirements he helped design years ago don't actually improve security very much. Put another way, when the situation changes, or new facts emerge, we must be willing to change as necessary with them.

Many of these problems indeed are preventable. But that's true only if the cybersecurity industry, and society as a whole, follows the lead of that

NIST researcher. We all must take a realistic look at the state of cybersecurity, admit the mistakes that have happened and change our thinking for the better. Only then can anyone—much less everyone—take on the task of devoting time, money and personnel to making the necessary changes for meaningful security improvements. It will take a long time, and will require inconvenience and hard work. But it's the only way forward.

12. Cybersecurity*

Protecting Court Data

BRIAN MCLAUGHLIN

In June 2016, the international activist hacker group Anonymous Legion claimed responsibility for a cyberattack on the State of Minnesota Judicial Branch's website. This was not the first attack on the Minnesota judiciary, who suffered two similar "Denial of Service" attacks in December 2015. Though the cyber attackers disrupted Minnesota's website functionality for 10 days, they did not breach any of protected data assets in court systems. Minnesota's experience is unfortunately not unique. Court systems, both state and federal, are guardians of sensitive information for individuals and organizations. This extraordinary responsibility makes them a ripe target for cyberattacks. To properly protect data assets, courts must coordinate internally, and with the executive and legislative branches, in cyberattack prevention and response.

The federal Judiciary designates cybersecurity as their top administrative priority. As Judge Julia Gibbons, chair of the United States Judicial Conference's Budget, states, "Judiciary systems have and will continue to be targeted, like government and commercial entities everywhere." A multitude of entry points exist for cyberattacks or cyber breaches on the judicial branch, including information systems, networks (including WiFi), employee personal devices (including smartphones) and an array of court technology. Beyond the damaging consequences of disrupting court operations, hackers target the trove of sealed information in judiciary systems. A sampling of this data includes:

---

*Originally published as Brian McLaughlin, "Cybersecurity: Protecting Court Data," *PA Times*, May 26, 2017. Reprinted with permission of the publisher.

- Personal identifiers—including social security numbers, credit cards, and bank accounts
- Confidential informant names and search warrants in criminal cases
- Records on cases involving children and families, including adoptions
- Trade secrets in civil cases involving businesses and corporations
- Victim data in domestic violence and sexual assault cases
- Grade jury records and testimony
- Medical and psychological reports

Types of Cyberattacks

The Federal Bureau of Investigation (FBI) defines a cyber incident as "a past, ongoing, or threatened intrusion, disruption, or other event that impairs or is likely to impair the confidentiality, integrity, or availability of electronic information, information systems, services, or networks." Four types of cyberattacks are particularly concerning for court officials:

- **Denial of Service (DoS) attacks**—This type of cyberattack prevents legitimate users from accessing network services. DoS attacks are most commonly effectuated by overwhelming servers with traffic to a specific site.
- **Phishing**—This type of attack uses social engineering to solicit personal information from unsuspecting users. Phishing e-mails appear to look legitimate, and ask users to enter items such as user names or passwords to compromise accounts.
- **Ransomware**—This type of cyberattack infects software and locks access to data until a ransom is paid. Through phishing e-mails, drive-by downloading and unpatched vulnerabilities, hackers attempt to extort users by encrypting their data until the prescribed conditions are met.
- **Spyware**—This type of cyberattack, also known as adware, infects a computer by producing pop-up ads, re-directing browsers and monitoring a user's internet activity.

Cyberattack Prevention

Cyberattack prevention is a critical component of a court's security planning. An effective cyber incident response plan can help law enforcement locate and apprehend the perpetrators. Court officials must identify court data assets and vulnerabilities. Once identified, IT staff can establish necessary layers of protection and documented protocols for managing systems. Regular testing of the cyberattack protection plan is essential, along with adjusting systems to emerging threats. IT staff must be attuned to the software updates and new technology in virus detection. In addition to planning, ongoing employee training is key to avoiding compromising activity.

Keeping pace with cyber criminals requires being on the cutting edge of security technology. Spending on security can save much more money in recovering assets. Consequently, the legislative branch also plays a vital role, as cybersecurity is increasingly a funding priority. In an era of challenges for public budgeting, courts must carefully tailor their budgetary requests.

Documented in recent Congressional testimony, the United States Courts methodically presented their request for $85 million to cover cybersecurity activities. And in their 2018–2019 budget request, the Minnesota Judicial Branch requested $1.968 million to expand its efforts to mitigate the risk of data breaches, data corruption, system outages, document/data loss and cyberattacks.

Cyberattack Response

Properly responding to a cyberattack requires immediate, strategic action. A cybersecurity incident response team should be set in the planning process. Key first steps include pinpointing the area of intrusion and scope of the attack. Once assessed, the attack should be reported to at least one law enforcement agency, and explained to the public.

Law enforcement agencies in the executive branch are a vital component in the response to a cyberattack. The United States Department of Justice provides best practices for responding to and reporting cyberattack incidents. The FBI adds clear guidelines for cyberattack incident reporting. Minnesota officials worked with the FBI Cyber Task Force in the aftermath of their DoS attacks. In the Department of Homeland Security, the United States Computer Emergency Response Team (U.S.-CERT) develops timely and actionable cybersecurity information for federal departments and agencies, as well as state and local governments.

Summary

Court systems are guardians of sensitive data for individuals and organizations. But they cannot fulfill this responsibility alone. In cyberattack prevention and response, the judicial branch needs the resources of the other branches of government to effectively protect data assets.

This chapter presents the personal views of the author, and does not represent the New Jersey Judiciary.

13. Health-Care Industry Increasingly Faces Cybersecurity Breaches*

MARGARET STEEN

The scenarios are chilling: A busy hospital suddenly cannot use any of its electronic medical records or other computerized systems. The victim of a ransomware attack, the hospital will not regain access without paying those who locked down the records—if at all.

At another hospital, hackers find a way to connect to the software that controls IV pumps, changing their settings so they no longer deliver the correct doses of medication.

Cybersecurity experts say these are among the situations they worry about when they consider the health-care industry—which, with its reliance on technology and a wealth of data, is increasingly a target of cybercrimes.

"We have seen in recent years an escalation in the risk to health-care organizations from cyberthreats," said Steve Curren, director of the Division of Resilience in the Office of Emergency Management, part of the U.S. Health and Human Services Department's Office of the Assistant Secretary for Preparedness and Response. "Since 2014, we have had 10 distinct breach incidents of health-care organizations where the breach resulted in the compromising of more than 1 million patient records."

And starting around 2016, attackers ramped up ransomware attacks against health-care systems. "That has been very disruptive," Curren said, sometimes forcing hospitals to implement emergency procedures.

---

*Originally published as Margaret Steen, "Health-Care Industry Increasingly Faces Cybersecurity Breaches," *Government Technology*, October 25, 2017. Reprinted with permission of the publisher.

Ransomware attacks have "impacted health care directly," said Monzy Merza, head of security research for Splunk, an enterprise software company. "There were several reports of UK hospitals unable to administer X-rays. The computer equipment attached to the X-ray machines was compromised and attacked by ransomware and rendered inoperable for some period of time."

Experts say there are a number of reasons for the increased risk—and challenges, some unique to health care, in mitigating it.

"Cybersecurity is somewhat of a nascent discipline," Merza said. "We're still learning. Manufacturers are learning how to operate in this new world. The same is true for the operators and owners of these technologies, who are also learning what the best practices are and how to manage them."

Reasons the Health-Care Industry Makes an Attractive Target for Cybercrimes

Lots of data. People launch cyberattacks for a variety of reasons, said Phyllis A. Schneck, managing director and global leader of cybersolutions for Promontory Financial Group, an IBM Company, and former chairman of the National Board of Directors of the FBI's InfraGard program. Some are simply having fun; others are deliberately trying to destroy infrastructure.

But a common reason is to steal intellectual property or personal information for financial gain. The health-care sector is "a resource-rich environment" for those looking for information due to the wealth of information health-care providers store: family history, medical history, financial information.

"There's a street value to people's personal information, and the health-care sector is an excellent source of it," Schneck said. Trade secrets can also be sold for profit.

Health-care organizations also have a lot of information that can be valuable to those who want to commit health insurance fraud, Medicare fraud or identity theft, Curren said.

Ransomware attacks are yet another way to make money.

"A lot of the bang for your buck is in locking up the system: Send in malware that freezes all the computers in the hospitals, then say, 'I'll send the code to unlock this if you send money,'" said Deborah A. Levy, a retired captain with the U.S. Public Health Service and currently professor and chair of the epidemiology department at the University of Nebraska Medical Center's College of Public Health. With the move toward electronic health records, the industry has become a bigger target.

Individual medical records may also be attractive if they include sensitive information about celebrities, for example, though in general there is less of a market for them.

Connections among diverse organizations. "The reason we're seeing more of this now is because of the connectivity of networks and devices to the network," Merza said. "There are clear advantages to connected devices—automation, information sharing, knowledge enrichment, contextualization. But with that network connectivity, you're opening yourself up to attack."

Organizations within the health-care sector also need to communicate with each other, so even if a large insurance company or hospital is able to secure its data, it may still be vulnerable when it shares connections with smaller organizations that have fewer resources for cybersecurity.

"We have a very diverse sector," Curren said, ranging from large health insurance organizations with a lot of resources to very small clinical practices.

An open culture. "Health care has an open, sharing culture—as is appropriate to support its primary mission—but this culture also complicates the issues of security and privacy," said the June 2017 *Report on Improving Cybersecurity in the Health Care Industry*, produced by the Health Care Industry Cybersecurity Task Force of the U.S. Department of Health and Human Services.

This means it has been harder for health-care organizations to secure their data than some other industries.

"They do not have really good security technologies and privacy policies in place," said Niam Yaraghi, a nonresident fellow with the Brookings Institution's Center for Technology Innovation and assistant professor of operations and information management at the University of Connecticut's School of Business. "They are like the only house in the very affluent neighborhood that doesn't have a security system."

"The first and foremost mission of every health-care organization is to cure the sick and help the patient," Yaraghi said. "If you're being rushed to the emergency department, the first thing in your mind is, 'I hope the physicians at this hospital are really good doctors.' Whether they're going to keep your blood pressure and drug allergies confidential—that's not the first thing you care about. They are in the business of providing medical care to patients; they are not in the business of technology."

Focus on Solutions

The results of a breach for everyone involved in the health-care industry—hospitals, clinics, researchers and patients—can range from annoying to catastrophic.

Patients could be harmed or even die. Many people—both patients and health-care workers—could be inconvenienced by systems going down. And bad publicity could harm clinics and hospitals in areas where consumers have choices.

"It's a competitive business—if a facility has gotten hit, that might influence where the public chooses to go," Levy said.

Prevention is the best solution—but it, too, poses challenges. Experts offer these ideas for shoring up security to prevent or mitigate attacks:

Education and awareness. "In the past, it was much more challenging implementing cybersecurity features because people didn't consider it a must," said Idan Edry, CEO of Trustifi LLC. "They said, 'I've never been hacked, nobody stole any of my information, so I'm fine.'"

Today, those on the front lines of using the more secure systems—including patients and medical professionals—are more aware of the importance of cybersecurity. Continued education will help ensure that the people who need to use the secure systems are on board.

Simplicity. The more complex a system is, the harder it can be to keep updated to guard against cyberattacks.

"Keep it simple: Don't have too many disparate things where if you make one update it breaks everything else," Schneck said. "The more hot, new devices that you have, the more openings you have."

Backup systems. When cybersecurity systems fail to prevent an attack, good backups can make it easier to recover.

"In the case of ransomware, it's important to have very good backups, so that when something is compromised, you're able to get back up and running," Merza said.

Emergency planning. Cybersecurity may be an emerging challenge, but emergency managers can tackle it by using strategies similar to those they use for other situations. "If a hospital gets disrupted by a cyberincident, it's the same as if it was disrupted by a water main break or a tornado or anything else," Curren said.

Constant vigilance. Both manufacturers and owners of devices bear some responsibility for preventing attacks. Users and operators should be prepared to follow best practices for installing and testing the updates.

"Start with the fundamentals," Merza said. Manufacturers should be constantly evaluating bugs and vulnerabilities of their equipment and sharing that information with owners. "How quickly can manufacturers identify the problem, come up with the fix and distribute the fix to the users of those devices?"

Realistic regulations. Cybersecurity plans need to keep in mind the mission and culture of the health-care industry.

For example, it's easy to say all operators should immediately install all patches. But "sometimes it is not feasible for any number of reasons," Merza said. Government agencies that regulate the systems may be slow with their approval. "The regulatory space is not equipped today to handle the evolving nature of threats and the speed with which technological development is happening.

There is an opportunity now for regulatory bodies to work with operators and manufacturers to understand the on-the-field requirements so people can implement them in a reasonable fashion."

Healthy attitude toward risk. It's easy to blame doctors for being reluctant to learn a new electronic medical record system, for example, or update their computers.

"Doctors are geniuses in how they figure out how to help people, but notorious for not being meticulous about cybersecurity," Schneck said.

But it is important for those in charge of cybersecurity to keep the true goals of everyone who uses the systems in mind. Researchers need to be able to share information and produce new drugs. Health-care providers need to be able to exchange patient information. Some security measures may make it hard for health-care professionals to do their jobs. The key is to consider cybersecurity through the lens of risk management, Schneck said.

"It's not the doctor's fault that he is too busy and he thinks that he doesn't have time for remembering a complicated password that cannot be hacked into, not the nurse's fault that she is under so much pressure that she cannot read every email very carefully and figure out that it's a phishing email," Yaraghi said. "I do not blame physicians and people in the health-care industry at all."

Cooperation. So many of the players in the health-care system are connected to each other—hospitals communicate with doctors' offices, pharmacies and insurance companies, for example—that an attack on one entity with weaker security could threaten others.

"There's a real strong sense developing in health care that we have to do this together, and we have to be committed to sharing information with one another to make this work," Curren said. For example, hospitals need to notify each other of attempted attacks so other hospitals can prevent them.

In addition, a long-term solution would be for device manufacturers to "develop products and services that are hard to compromise," Merza said. "The government, the manufacturers and the operators of these devices all really have to work together in the best interests of the public health-care population."

14. It's in the Mail*

Aetna Agrees to $17M Payout in HIV Privacy Breach

ELANA GORDON

Aetna settled a lawsuit for $17 million Wednesday over a data breach that happened in the summer of 2017. The privacy of as many as 12,000 people insured by Aetna was compromised in a very low-tech way: The fact that they had been taking HIV drugs was revealed through the clear window of the envelope.

"I was shocked," said Sam, who distinctly recalls the day he received the notice in August. (*Kaiser Health News* and NPR agreed not to use his full name because he worries about how going public with his HIV status might affect his work.) The letter came to his mailbox in an apartment complex in New Jersey. He wasn't directly involved in the lawsuit but says the letter hit a level of vulnerability he had never felt before.

"I haven't disclosed my HIV status to my parents," said Sam, 36, who is a civil rights attorney. "Let's say that letter had gotten forwarded to their house and someone happened to open the mail. Those were the types of things going through my mind."

In a statement, Aetna wrote: "Through our outreach efforts, immediate relief program and this settlement we have worked to address the potential impact to members following this unfortunate incident."

The insurer also said it is "implementing measures designed to ensure something like this does not happen again as part of our commitment to best practices in protecting sensitive health information."

---

*Originally published as Elana Gordon, "It's in the Mail: Aetna Agrees to $17M Payout in HIV Privacy Breach," *Kaiser Health News*, January 18, 2018. Reprinted with permission of the publisher.

In an ironic twist, the letters were sent in response to a settlement over previous privacy violation concerns. Aetna had required members to obtain HIV medications through mail-order pharmacies. The affected people had taken medication to treat HIV or to lower the risk of becoming infected with the virus, an approach called PrEP, or pre-exposure prophylaxis.

Lawsuits filed in 2014 and 2015 alleged that policy was discriminatory, that it prevented patients taking HIV medicine from receiving in-person counseling from a pharmacist and that it jeopardized members' privacy.

Aetna settled with the individual plaintiffs, changed its policy to allow members to fill HIV prescriptions in person at retail pharmacies, and, in turn, sent out notification letters to anyone who had filled prescriptions for HIV medications.

It was those notification letters that contained a large envelope window that exposed sensitive HIV information.

While the stigma surrounding HIV may be less severe than it used to be and treatments have improved greatly, Ronda Goldfein, director of the AIDS Law Project of Pennsylvania, said the reality is that serious discrimination still exists. That means protecting patient confidentiality is critical to ensuring people feel safe getting care.

As hundreds of calls from people who received the Aetna letter started coming into Goldfein's office and others around the country, she learned of more harrowing and devastating experiences. She said she heard from one man who had homophobic slurs painted on his door when neighbors saw the letter.

Other letter recipients felt the need to move out of their neighborhoods. For one woman, whose status became known in her tight-knit immigrant community, "she stopped being able to function, she stopped being able to go to work, and she lost her job," Goldfein said.

The AIDS Law Project of Pennsylvania and the Legal Action Center initially issued a demand letter in late August that the insurer stop the mailings. The company responded, setting up a relief fund for affected people and apologizing. "This type of mistake is unacceptable, and we are undertaking a full review of our processes to ensure something like this never happens again," the health insurer said.

Goldfein and others soon discovered that the mailing was more widespread than first thought: Up to 12,000 people had received it. Her agency, the Legal Action Center and Berger & Montague PC filed a lawsuit and sought class-action status.

The privacy breach as outlined in the proposed settlement was twofold: Aetna released the names of 13,480 people to its legal counsel and a vendor without proper authorization. Of those, 11,875 got the letter that revealed they were taking HIV medication.

The proposed settlement is awaiting approval in federal court, but in it Aetna has agreed to pay $17 million and set up new "best practices" to prevent something like this from happening again.

As part of the payout, the law firms are setting aside at least $12 million for payments of at least $500 to the estimated 11,875 people who may have received a letter exposing that information, acknowledging that "the harm was in the status being disclosed," Goldfein said. Plus, people won't have to file additional paperwork and go through more mailings pertaining to their HIV medications.

A fund will be set up for those who experienced additional financial or emotional distress. Individuals will be able to claim up to $20,000. The rest of the money will go toward legal fees and costs.

"It's a much bigger settlement than ordinary identity theft scenarios, where an online database has been breached and the main injury people are claiming is that they might be victims of identity theft and maybe have their financial information compromised," said William McGeveran, a specialist in privacy law and data breaches at the University of Minnesota.

The amount may be unusual, but McGeveran also said low-level breaches like this aren't. Companies may be so focused on IT security that they overlook other ways that privacy can be breached.

"They're more common than people realize," McGeveran said. "There's so much attention to cybersecurity, and rightly so, but a lot of medical privacy concerns are much more analog than that. They're about things being overheard, they're about paper records and in this case it's about a paper mailing."

Beyond the payout itself, she hopes the suit helps change the culture of companies when it comes to the attention paid to medical privacy, and the rights of people with HIV in particular. To highlight that, lawyers used "Andrew Beckett" as the pseudonym for the original plaintiff in the case, a Pennsylvania man from Bucks County.

It's a nod to the Tom Hanks character in the 1993 film "Philadelphia," who was fired after his law firm found out he had HIV. This "Beckett" is taking PrEP.

"HIV still has a negative stigma associated with it, and I am pleased that this encouraging agreement with Aetna shows that HIV-related information warrants special care," the man known as Beckett said in statement.

This story is part of a partnership that includes WHYY, NPR and Kaiser Health News.

15. Explainer*

What Is Hacking?

JAMES H. HAMLYN-HARRIS

Last week, we woke to news that the largest cyber attack ever was underway in Europe, with reports of global internet speeds falling as a result of an assault on the anti-spamming company Spamhaus.

In recent weeks, the Reserve Bank of Australia has been the target of a cyber attack, as have South Korean banks and broadcasters and BBC Twitter accounts.

The above stories were all reported as "hacking"—a blanket term readily used to encompass a whole range of attacks, from crashing a server to more sophisticated infiltration, such as stealing passwords. But, generally, news stories don't discriminate.

So what are hackers and their methods really like? What follows is something of a glossary, to cut out (or at least bookmark) and keep.

Types of Hackers

Phreakers: Perhaps the oldest type of computer hackers, Phreakers discover how telephone systems work and use their knowledge to make free phone calls.

In the past, phone phreakers used what we now think of as hacking techniques to access mainframe computers and programmable telephone switches to obtain information, alter records or evade capture.

*Originally published as James H. Hamlyn-Harris, "Explainer: What Is Hacking?," *The Conversation*, April 1, 2013. Reprinted with permission of the publisher.

Famous (and now retired) phreakers include Kevin Mitnick, Kevin Poulsen and Apple founders Steve Jobs and Steve Wozniak.

Crackers: These guys bypass (crack) security controls on proprietary software, DVDs, computer games and Digital Rights Management (DRM)–protected media.

Crackers trade, share and publish game "cracks," patches, serial numbers and keygens (activation key generators). They also embed malware in their cracks and patches forming Trojans to deter outsiders (mostly "script kiddies"; see below) from using their code.

Unsuspecting people who use their cracks more often than not find themselves infected with worms and viruses (explained below). Such infections often bypass anti-virus tools and firewalls, and are probably responsible for most of the malware on teenagers' home computers.

Black Hat Hackers: These are crackers who actively develop malware and intrusion techniques and tools for evil purposes, Black Hats are motivated by profit.

Criminal organizations, foreign governments and spy agencies will pay handsomely for the latest zero-day (not publicly known) exploit.

Journalist Brian Krebs recently reported a bidding war for a Java exploit valued at more than U.S. $5,000.

White Hat Hackers: These are the good guys. White Hats, also known as "ethical hackers" and "pen-testers," are security researchers.

They test systems (often using the same tools as Black Hats, but within the law) by conducting penetration testing and security audits as a service for businesses and organizations that don't want to be hacked.

White Hats report on any vulnerabilities found and what needs to be done to fix them. Both the U.S. and Australian governments have set up competitions to encourage school and university students to take up (White Hat) hacking as a career.

(My Swinburne team competed in the pilot version of Australia's Cyber Challenge in 2012 and scored higher than all other Victorian universities.)

Grey Hat Hackers: Grey Hats generally work within the law but may publish vulnerabilities and exploits or sell exploits to unknown buyers without asking too many questions.

They may also report vulnerabilities to software vendors anonymously to avoid prosecution. Unfortunately some vendors object to having their defective code discovered and discourage security research on their products.

Script kiddies: Also known as "skiddies," these are a growing number of amateur Black Hats who cannot develop their own code but can adapt other people's exploits and use hack tools to attack organizations and each other.

Script kiddies find the term offensive and have been known to launch cyber-attacks against people who have denigrated them or their skills.

It is likely that many of the "hackers" associated with online protest group Anonymous are script kiddies.

Cyber-troops, cyber-soldiers: These are state-sponsored military personnel trained in hacking techniques who use malware and hacking techniques to spy, gather intelligence, steal intellectual property and disrupt enemy systems.

Spammers and Phishers: Spammers use programs—spambots—to automatically send email, SMSs, instant messages and tweets to potential buyers of their products.

Phishers use the same technologies (and fake "pharming" sites) to entice victims to click on links (and type in user-names and passwords) and download and install malware. The book Spam Kings recounts the early history of many spammers.

Types of Hacks

Now that we know who the bad guys are, let's consider what they do and how their actions are likely to affect people.

Script injection (SQL, JavaScript) attacks: Most websites are connected to databases. With Structured Query Language (SQL) injection, attackers run their own code on these databases, allowing them to change records, delete data and extract private information such as credit card numbers, passwords or password hashes.

JavaScript injection happens through publicly-writable web sites such as Facebook, Twitter and sites with forums and discussion boards. If not properly filtered, an attacker can upload script that extracts private information from people visiting the site.

Scripts can bypass firewalls to extract user credentials, track user activities, install malware and even turn on the web camera and microphone. The simplest way to prevent such attacks is to turn off scripting (in your browser).

The Firefox NoScript plug-in is an easy way to do this.

Password cracking: Simply put, if an attacker can guess your password, he or she can take over your computer. Most computer users are overwhelmed by the number of account names and passwords they have to remember, so they tend to re-use them.

An attacker can use SQL injection to recover passwords or password hashes from a poorly-secured website, and then try the same user-names and passwords to log into high-value sites such as bank accounts.

Websites and email systems that restrict password length are the easiest to attack.

Brute force attacks: These use automated tools to guess the password or re-create the password hash.

The most effective ways of preventing this is to (a) use long passwords, and (b) use different passwords.

DoS/DDoS: (Distributed) Denial of service attacks are generally launched against organizations, whose servers are flooded with "broken" network communications that cause the servers to slow down or even crash.

Companies that rely on online trading will lose a lot of money (and - reputation) if this happens, and will often pay the attackers to call off the attack.

Viruses, worms and Trojans: These are infection carriers used to distribute malware. Viruses travel by thumb drives, worms travel through the internet, and Trojans are downloaded by unsuspecting users.

Anti-virus software will stop most of this, but not the latest (or zero-day) malware attacks.

Crimeware, hijackers and ransomware: Black Hat hacking has matured into an industry. Hackers can purchase crimeware packs for a few thousand dollars and start up a business distributing malware, accepting payments and laundering money.

Hijackers take over your web browser and redirect you to advertising sites. Ransomware infects your computer and prompts you to call a toll-free number, where you can pay to have your computer remotely "disinfected."

Man-in-the-browser malware, such as Zeus, can intercept your online banking sessions in your browser and phone, draining your account by sending money to the attackers.

Bots and bot-nets: Bots emulate human users. Once a bot has infected your computer, you are "owned." Your computer (now a zombie) is remotely controlled by a bot herder who can use it and hundreds of thousands of other zombies to launch DDoS attacks, crack passwords, send spam and host illegal content.

Protect Yourself

We can only minimize the risks, but the risks are well understood. Turn off scripting, maintain your anti-virus, don't read unsolicited emails, use long passwords, use different passwords, don't download programs you didn't go looking for, be skeptical … and finally: learn about computer security (to find out what else you can do).

There's no need to be paranoid. Just be careful. White Hat hackers are there to help by exposing the risks and testing the systems. Trust them. They're the good guys.

16. How the Chinese Cyberthreat Has Evolved*

Dorothy Denning

With more than half of its 1.4 billion people online, the world's most populous country is home to a slew of cyberspies and hackers. Indeed, China has likely stolen more secrets from businesses and governments than any other country.

Covert espionage is the main Chinese cyberthreat to the U.S. While disruptive cyberattacks occasionally come from China, those that cause overt damage, like destroying data or causing power outages, are more common from the other top state threats, namely Russia, Iran and North Korea.

But Chinese cyberaggression toward the U.S. has been evolving. Before their espionage became a serious threat, Chinese hackers were conducting disruptive cyberattacks against the U.S. and other countries.

Hackers Unite

Chinese hackers were among the first to come together in defense of their country. Their first operation against the U.S. occurred in 1999 during the Kosovo conflict, when the U.S. inadvertently bombed the Chinese embassy in Belgrade, killing three Chinese reporters. The patriotic hackers planted messages denouncing "NATO's brutal action" on several U.S. government websites.

Chinese hackers struck the U.S. again in 2001 after a Chinese fighter plane collided with a U.S. reconnaissance aircraft. The midair collision killed

---

*Originally published as Dorothy Denning, "How the Chinese Cyberthreat Has Evolved," *The Conversation*, October 4, 2017. Reprinted with permission of the publisher.

the Chinese pilot and led to the forced landing and detention of the American crew. Both Chinese and American hackers responded with disruptive cyberattacks, with the Chinese hackers defacing thousands of U.S.-based websites, including the White House site.

What is especially important about this incident, though, is what happened next. *The People's Daily*, China's Communist Party newspaper, issued an editorial decrying the attack against the White House. The paper called it, and the other attacks, "web terrorism" and "unforgivable acts violating the law." On the anniversary of the incident in 2002, the government asked Chinese hackers to forgo further attacks against U.S.-based sites. They complied.

That was the last big cyberattack from Chinese patriotic hackers against the U.S. While Russia seems to condone, if not outright encourage or even sponsor, its patriotic hackers, China has taken a stance against that sort of activity, at least with respect to U.S.-based sites.

Targets at Home

In addition to reining in its patriotic hackers, China appears to have refrained from conducting cyberattacks that cause overt damage to critical infrastructure in other countries, like the Russians did to Ukraine's power grid. However, it has used disruptive cyberattacks to help enforce censorship policies within its own borders.

The Chinese government's "Great Firewall" keeps internet users in China from accessing censored foreign sites such as those that advocate Tibetan autonomy. Users' traffic is filtered based on domain names, internet addresses and keywords in web addresses.

Chinese hackers have also used denial-of-service attacks to temporarily take out sites whose activity the government wants to block. These attacks overwhelm target servers with large amounts of activity, preventing others from using the sites and often knocking the servers offline.

Back in 1999, the government launched DoS attacks against foreign websites associated with Falun Gong, a spiritual movement banned in China. Then in 2011, a Chinese military TV program showed software tools being used in possible cyberattacks against Falun Gong sites in the U.S. The tools were developed by the Electrical Engineering University of China's armed forces, the People's Liberation Army.

More recently, in 2015, U.S. and other foreign users visiting sites running analytics software from the Chinese search engine provider Baidu unwittingly picked up malware. The malicious code was injected into traffic going back to the users by a device collocated with the Great Firewall. The malware then launched DDoS attacks against GreatFire.org, a site that helps Chinese users evade censorship, and the Chinese language edition of the *New York Times*.

Espionage at the Forefront

By 2003, China's interest in cyberespionage was apparent: A series of cyberintrusions that U.S. investigators code-named "Titan Rain" was traced back to computers in southern China. The hackers, believed by some to be from the Chinese army, had invaded and stolen sensitive data from computers belonging to the U.S. Department of Defense, defense contractors and other government agencies.

Titan Rain was followed by a rash of espionage incidents that originated in China and were given code names like "Byzantine Hades," "GhostNet" and "Aurora." The thieves were after a wide range of data.

They stole intellectual property, including Google's source code and designs for weapons systems. They took government secrets, including user names and passwords. And they compromised data associated with Chinese human rights activists, including their email messages. Typically, the intrusions started with spear-phishing.

In 2013, the American cyberintelligence firm Mandiant, now part of FireEye, issued a landmark report on a Chinese espionage group it named "Advanced Persistent Threat 1." According to the report, APT1 had stolen hundreds of terabytes of data from at least 141 organizations since 2006.

The Mandiant report gave details of the operations and provided evidence linking those thefts to Unit 61398 of the People's Liberation Army—and named five officers of the unit. This was the first time any security firm had publicly disclosed data tying a cyberoperation against the U.S. to a foreign government. In 2014, the U.S. indicted the five Chinese officers for computer hacking and economic espionage.

Mandiant described APT1 as "one of more than 20 APT groups with origins in China." Many of these are believed to be associated with the government. A report from the nonprofit Institute for Critical Infrastructure Technology describes 15 state-sponsored advanced persistent threat groups, including APT1 and two others associated with PLA units. The report does not identify sponsors for the remaining groups.

The Five-Year Plan

According to the institute, China's espionage supports the country's 13th Five-Year Plan (covering the years 2016 to 2020), which calls for technology innovations and socioeconomic reforms. The goal is "innovative, coordinated, green, open and inclusive growth." The ICIT report said most of the technology needed to realize the plan will likely be acquired by stealing trade secrets from companies in other countries.

In its *2015 Global Threat Report*, the American cyberintelligence firm CrowdStrike identified dozens of Chinese adversaries targeting business sectors that are key to the Five-Year Plan. It found 28 groups going after defense and law enforcement systems alone. Other sectors victimized worldwide included energy, transportation, government, technology, health care, finance, telecommunications, media, manufacturing and agriculture.

China's theft of military and trade secrets has been so rampant that editorial cartoonists Jeff Parker and Dave Granlund depicted it as "Chinese takeout."

U.S.-China Agreement

In September 2015, President Obama met with China's President Xi Jinping to address a range of issues affecting the two countries. With respect to economic espionage, they agreed that their governments would not conduct or knowingly support cyber-enabled theft of business secrets that would provide competitive advantage to their commercial sectors. They did not agree to restrict government espionage, a practice that countries generally consider to be fair game.

In June 2016, FireEye reported that since 2014 there had been a dramatic drop in cyberespionage from 72 suspected China-based groups. FireEye attributed the reduction to several "factors including President Xi's military and political initiatives, the widespread exposure of Chinese cyberoperations, and mounting pressure from the U.S. Government." The ICIT believes China may also be asserting greater control over its operatives and focusing on unspecified high-priority targets.

The U.S.-China agreement also calls for the two countries to cooperate in fighting cybercrime. Just weeks after the deal was signed, China announced it had arrested hackers connected with the 2015 intrusions into the Office of Personnel Management's database. Those had exposed highly sensitive personal and financial data of about 22 million federal employees seeking security clearances. The *Washington Post* observed that the arrests could "mark the first measure of accountability for what has been characterized as one of the most devastating breaches of U.S. government data in history."

The cyberthreat to the U.S. from China is mostly one of espionage, and even that threat seems to be declining. Nevertheless, companies need to be wary of losing their data, not just to China, but to any country or group seeking to profit from U.S. trade secrets and other sensitive data. That calls for staying ahead of the cybersecurity curve.

17. Now That Russia Has Apparently Hacked America's Grid, Shoring Up Security Is More Important Than Ever*

THEODORE J. KURY

Hackers taking down the U.S. electricity grid may sound like a plot ripped from a Bruce Willis action movie, but the Department of Homeland Security and the FBI recently disclosed that Russia has infiltrated "critical infrastructure" like American power plants, water facilities and gas pipelines.

This hacking is similar to the 2015 and 2016 attacks on Ukraine's grid. While it hasn't risen beyond scouting mode, the specter of sabotage in the U.S. now seems more realistic than it used to.

Clearly, there's no time to waste in shoring up the grid's security. Yet getting that done is not easy, as I've learned through my research regarding efforts in to stave off outages in hurricane-prone Florida.

A Catch-22

There is no way to completely protect the grid. Even if that were possible, utilities tend to adopt new and better security procedures after mishaps, boosting the chance that some attacks will succeed.

---

*Originally published as Theodore J. Kury, "Now That Russia Has Apparently Hacked America's Grid, Shoring Up Security Is More Important Than Ever," *The Conversation*, April 11, 2018. Reprinted with permission of the publisher.

Regulation at the state and federal levels makes it hard for utilities and regulators to work together to get this job done.

Utilities can charge their customers only what it takes for them to cover reasonable expenses. Regulators must approve their rates through a process that needs to be open to public scrutiny.

Say, for example, a power company is building a substation. The utility would disclose what it spent on construction, prove that it picked its contractors responsibly and explain how this new capacity is enhancing its service. The regulator then must decide what rate hikes, if any, would be reasonable—after hearing out everyone with something at stake.

Following this routine is harder with cyberdefense spending. Security concerns make it tough if not impossible for utilities to say what they're doing with that money. Regulators, therefore, have a hard time figuring out whether utilities are spending too much or too little or maybe even wasting money on an unnecessary expense.

If regulators blindly approve these rate hikes, it can be an abdication of their duties. If they reject them, utilities get penalized for shoring up their security and then lose an incentive to keep doing the right thing.

To Err Is Human

Even though the idiosyncrasies of utility regulation make cyberdefense a more complicated issue than it might otherwise be, tools to manage this risk are available.

Mitigating the damage that human error can cause in response to malicious attacks, for example, may not demand any spending beyond what it costs to teach workers at utilities and their contractors to refrain from blindly opening perilous email attachments, the avenue into the electricity system used by hackers in the 2015 Ukraine attacks and in the system breaches the government recently disclosed.

Indeed, hackers delivered almost 94 percent of all malware in 2016 through email systems. Clearly, more widespread awareness of the need to keep an eye out for phishing attacks will help secure infrastructure.

Regulators have been studying strategies that might enhance cybersecurity. Standards are already in place in the U.S., Canada and part of Mexico for utilities to assess their capability to prevent or detect cyberattacks.

Preventive measures can include states adopting new regulations that protect utilities' confidential information and doing more to train utility workers to spot and confront cybersecurity threats.

It's also important that regulators recognize that securing systems is an ongoing process. It can never really end because as system security measures change, hackers devise new ways to circumvent them.

Grid Resilience

Grid resilience strategies can help to maintain service regardless of the source of the outage. For example, many utilities have invested in "self-healing" systems that isolate glitches in the grid and quickly restore service amid outages.

Here's an example of how that works. During Hurricane Matthew in Florida, in 2016, Florida Power and Light identified a threatened substation and isolated it from the rest of the grid. This measure protected its customers by ensuring that outages at that substation would not spread.

Utilities can also create microgrids, or portions of the grid that can be isolated from the rest of the system in the event of an attack. Most of these systems have been designed to improve resilience in the event of natural disasters and storm events. But they can help defend the grid against cyberattacks as well.

Public concerns over grid security are more justified than ever. But I believe that minimizing the risk of a catastrophic infrastructure attack is within reach. All it will take is for utilities to educate their workers on system security while the government updates its rules and practices—and for everyone involved to keep doing what they can to avert outages of all kinds and to restore power as quickly as possible when outages occur despite those efforts.

18. Ransomware Victims Urged to Report Infections to Federal Law Enforcement*

FEDERAL BUREAU OF INVESTIGATION

The FBI urges victims to report ransomware incidents to federal law enforcement to help us gain a more comprehensive view of the current threat and its impact on U.S. victims.

What Is Ransomware?

Ransomware is a type of malware installed on a computer or server that encrypts the files, making them inaccessible until a specified ransom is paid. Ransomware is typically installed when a user clicks on a malicious link, opens a file in an e-mail that installs the malware, or through drive-by downloads (which does not require user-initiation) from a compromised Web site.

Why We Need Your Help

New ransomware variants are emerging regularly. Cyber security companies reported that in the first several months of 2016, global ransomware infections were at an all-time high. Within the first weeks of its release, one particular ransomware variant compromised an estimated 100,000 computers a day.

---

*Public document originally published as Federal Bureau of Investigation, "Ransomware Victims Urged to Report Infections to Federal Law Enforcement," https://www.ic3.gov/media/2016/160915.aspx (September 15, 2016).

Ransomware infections impact individual users and businesses regardless of size or industry by causing service disruptions, financial loss, and in some cases, permanent loss of valuable data. While ransomware infection statistics are often highlighted in the media and by computer security companies, it has been challenging for the FBI to ascertain the true number of ransomware victims as many infections go unreported to law enforcement.

Victims may not report to law enforcement for a number of reasons, including concerns over not knowing where and to whom to report; not feeling their loss warrants law enforcement attention; concerns over privacy, business reputation, or regulatory data breach reporting requirements; or embarrassment. Additionally, those who resolve the issue internally either by paying the ransom or by restoring their files from back-ups may not feel a need to contact law enforcement.

The FBI is urging victims to report ransomware incidents regardless of the outcome. Victim reporting provides law enforcement with a greater understanding of the threat, provides justification for ransomware investigations, and contributes relevant information to ongoing ransomware cases. Knowing more about victims and their experiences with ransomware will help the FBI to determine who is behind the attacks and how they are identifying or targeting victims.

Threats to Users

All ransomware variants pose a threat to individual users and businesses. Recent variants have targeted and compromised vulnerable business servers (rather than individual users) to identify and target hosts, thereby multiplying the number of potential infected servers and devices on a network. Actors engaging in this targeting strategy are also charging ransoms based on the number of host (or servers) infected. Additionally, recent victims who have been infected with these types of ransomware variants have not been provided the decryption keys for all their files after paying the ransom, and some have been extorted for even more money after payment.

This recent technique of targeting host servers and systems could translate into victims paying more to get their decryption keys, a prolonged recovery time, and the possibility that victims will not obtain full decryption of their files.

What to Report to Law Enforcement

The FBI is requesting victims reach out to their local FBI office and/or file a complaint with the Internet Crime Complaint Center, at www.IC3.gov, with the following ransomware infection details (as applicable):

1. **Date of Infection**
2. **Ransomware Variant** (identified on the ransom page or by the encrypted file extension)
3. **Victim Company Information** (industry type, business size, etc.)
4. **How the Infection Occurred** (link in e-mail, browsing the Internet, etc.)
5. **Requested Ransom Amount**
6. **Actor's Bitcoin Wallet Address** (may be listed on the ransom page)
7. **Ransom Amount Paid** (if any)
8. **Overall Losses Associated with a Ransomware Infection** (including the ransom amount)
9. **Victim Impact Statement**

The Ransom

The FBI does not support paying a ransom to the adversary. Paying a ransom does not guarantee the victim will regain access to their data; in fact, some individuals or organizations are never provided with decryption keys after paying a ransom. Paying a ransom emboldens the adversary to target other victims for profit, and could provide incentive for other criminals to engage in similar illicit activities for financial gain. While the FBI does not support paying a ransom, it recognizes executives, when faced with inoperability issues, will evaluate all options to protect their shareholders, employees, and customers.

Defense

The FBI recommends users consider implementing the following prevention and continuity measures to lessen the risk of a successful ransomware attack.

- Regularly back up data and verify the integrity of those backups. Backups are critical in ransomware incidents; if you are infected, backups may be the best way to recover your critical data.
- Secure your backups. Ensure backups are not connected to the computers and networks they are backing up. Examples might include securing backups in the cloud or physically storing them offline. It should be noted, some instances of ransomware have the

capability to lock cloud-based backups when systems continuously back up in real-time, also known as persistent synchronization.
- Scrutinize links contained in e-mails and do not open attachments included in unsolicited e-mails.
- Only download software—especially free software—from sites you know and trust. When possible, verify the integrity of the software through a digital signature prior to execution.
- Ensure application patches for the operating system, software, and firmware are up to date, including Adobe Flash, Java, Web browsers, etc.
- Ensure anti-virus and anti-malware solutions are set to automatically update and regular scans are conducted.
- Disable macro scripts from files transmitted via e-mail. Consider using Office Viewer software to open Microsoft Office files transmitted via e-mail instead of full Office Suite applications.
- Implement software restrictions or other controls to prevent the execution of programs in common ransomware locations, such as temporary folders supporting popular Internet browsers, or compression/decompression programs, including those located in the AppData/LocalAppData folder.

Additional considerations for businesses include the following:

- Focus on awareness and training. Because end users are often targeted, employees should be made aware of the threat of ransomware, how it is delivered, and trained on information security principles and techniques.
- Patch all endpoint device operating systems, software, and firmware as vulnerabilities are discovered. This precaution can be made easier through a centralized patch management system.
- Manage the use of privileged accounts by implementing the principle of least privilege. No users should be assigned administrative access unless absolutely needed. Those with a need for administrator accounts should only use them when necessary; they should operate with standard user accounts at all other times.
- Configure access controls with least privilege in mind. If a user only needs to read specific files, he or she should not have write access to those files, directories, or shares.
- Use virtualized environments to execute operating system environments or specific programs.
- Categorize data based on organizational value, and implement physical/logical separation of networks and data for different organizational units. For example, sensitive research or business

data should not reside on the same server and/or network segment as an organization's e-mail environment.

- Require user interaction for end user applications communicating with Web sites uncategorized by the network proxy or firewall. Examples include requiring users to type in information or enter a password when the system communicates with an uncategorized Web site.
- Implement application whitelisting. Only allow systems to execute programs known and permitted by security policy.

19. Ransomware Attacks Illustrate the Vulnerabilities That Local Government Entities Face*

MARY SCOTT NABERS

The media has focused on cyberattacks related to election systems lately. Many individuals, public officials, and company executives, however, have been just as worried about ransomware attacks.

Ransomware attacks infect computer networks with a virus that totally shuts down a computer or a network. It prevents access and demands payment to release and restore data on the machine or network. Recent examples of ransomware attacks illustrate the vulnerabilities that government entities face. The ransom costs are exorbitant while the risk of either a loss of data or a service outage is terrifying.

President Trump's proposed 2018 budget increases cybersecurity personnel across multiple key agencies and, if passed, it will boost the Department of Homeland Security (DHS) cybersecurity unit budget to nearly $3.3 billion. That's all good, but most of the recent ransomware attacks have occurred at the local levels of government and the federal government allocates no funding for that problem. According to a recent report, states spend between 0 percent and 2 percent of their IT budgets on cybersecurity—while all best practices suggest that spending should represent between 10 percent and 15 percent of an organization's budget.

*Originally published as Mary Scott Nabers, "Ransomware Attacks Illustrate the Vulnerabilities That Local Government Entities Face," ICMA Blog, International City/County Management Association (July 12, 2017). Reprinted with permission of the publisher.

In April, a hacker infected computers at the city of Newark, N.J. The virus rendered all machines unusable. The city's network was compromised and it disrupted digital services. The hacker demanded $30,000 in Bitcoin, an Internet currency that is difficult to trace. Most ransomware encrypts common computer files and requires a password to unlock them.

Last November, the San Francisco Municipal Transportation Agency (SFMTA), which operates the MUNI light rail system, was also attacked by ransomware. While the ransomware did not penetrate the agency's network, it did shut down ticket vending machines. Hackers demanded $70,000 in Bitcoin. To prevent disruption in service, SF Muni offered free rides until the fare machines were operational again. The agency did not pay the ransom but the attack was very costly.

In August 2016, the city of Sarasota in Florida had a ransomware attack that shut its computer systems down by a type of ransomware that entered the city's system through a virus that was sent to one employee. Despite demands by hackers, the city did not pay the ransom and was finally able to recover its files. The cost of resources, lost productivity and inability to provide services, however, was very high.

Last February, the city of Los Angeles Integrated Security Operations Center (ISOC) identified 16 ransomware attacks in five city departments. The attacks were segmented, no data was lost, and no ransom was paid. But, analysts' biggest worry now is ransomware and they struggle to stay two steps ahead of these types of attacks. The city's proposed FY 2018 budget includes $2.25 million in funding to support cybersecurity initiatives.

The state of New York's proposed 2018 budget funds the creation of a new Cyber Incident Response Team, which will not only support state agencies, but also local governments, critical infrastructure statewide and schools. The team will provide outreach services and coordinated exercises and act as a first responder to reported cyber incidents.

Many state and local government leaders are committed to having trained personnel in-house so that cyberattacks do not represent the threats that are prevalent today. That will take time, resources and funding that most had not planned for in future budgets. The world has become a more frightening place and government networks are attractive targets for hackers, cyber sleuths and professionals. Taxpayers will ultimately pay much, if not all, of the cost of keeping the government's data safe from cyberattacks in the future. That is not a positive thought, but even less positive is the reality that individuals are just as vulnerable because personal computer systems are also a major target for ransomware attacks. Protection is available, but costly to citizens, just as it is to the government.

20. Ransomware Attacks on the Rise in 2017*

MARY ANN BARTON

Facing the loss of its data, officials in Montgomery County, Ala., authorized funds last week to pay a ransom to hackers to get its government back up and running. After the county's computer system was hit Sept. 19 by a ransomware attack, one of its options was to pay the ransom within seven days before data was destroyed. The county ended up paying between $40,000 to $50,000 to obtain nine bitcoins to pay the ransom, County Commission Chair Elton Dean said in a news conference. Dean said the loss of files would have cost the county about $5 million.

The county, which counts about 230,000 residents, was unable to issue vehicle tags or registrations or handle business or marriage license requests while it was tied down. The county's chief IT officer, Lou Ialacci, said all of the county's departments were affected.

Montgomery County isn't alone. There have been hundreds of ransomware victims this year and the FBI says the practice is on the rise. Ransomware "is a very big problem and it has not abated as yet," said Ron Yearwood, section chief for the FBI's Cyber Operations, headquartered in New York. When there's a major ransomware attack, the FBI's little known Cyber Action Team gets into the picture. "They're considered the elite among intrusion investigators," Yearwood said.

In the first quarter of 2017, the most recent figures available, there have been 745 victims of ransomware, losing more than $512,000 to cyber hackers, the FBI said, along with much more lost in work hours, etc. At that pace, the FBI could see more victims than last year, when 2,673 notified the crime-

*Originally published as Mary Ann Barton, "Ransomware Attacks on the Rise in 2017," *NACo County News*, October 2, 2017. Reprinted with permission of the publisher.

fighting agency about ransomware attacks. And those are only the attacks the FBI knows about. "Typically, we see under-reporting," Yearwood said.

While most county IT officials probably are aware that ransomware attacks are normally delivered through spam emails or "spear phishing emails," which target specific individuals, in newer instances of ransomware, some cyber criminals aren't using emails at all, according to the FBI. They can bypass the need for an individual to click on a link by seeding legitimate websites with malicious code, taking advantage of unpatched software on end-user computers, the FBI warns.

How do you keep the bad guys out? Yearwood said that some of the best ways to prevent a ransomware attack include two-factor authentication, limiting remote access and segregating critical data behind multiple defenses.

A little-known defense that is outside of traditional thinking: "We talk about looking outward" to see if a hacker is getting into your system, Yearwood said. "If an adversary is able to get past your defenses without your knowledge, and the outwardly looking defenses don't catch them or alert to them ... the intrusion goes unnoticed, it doesn't get caught in that capacity and they can be on the network for a very long and extended period of time. So, I would challenge any potential victims out there to not just protect the perimeter, but also look at the traffic going across their network, do some auditing on their system. An example of that would be passive GMS monitoring. It could help you identify suspicious outbound connections."

If you've been hacked, don't touch anything until you've contacted the FBI, Yearwood said. If you are contacting the FBI about a possible hack, pick up the phone and call them, don't try to contact them via email on the computer system, he advised.

If you shut down or disconnect your system, it could make matters worse.

And be sure to establish a relationship with your local FBI office before you need them, Yearwood noted.

The FBI says it does not recommend paying a ransom in response to a ransomware attack. Paying a ransom not only doesn't guarantee that you will get your data back—there have been instances where organizations never got a "decryption key" after having paid the ransom, the agency said. Paying a ransom could embolden the cyber criminals to target more organizations and offers an incentive for other criminals to get involved in this type of illegal activity, the FBI said. And by paying a ransom, an organization might inadvertently be funding other illicit activity associated with criminals.

But "each incident is different," Yearwood said. "Each will have to determine the best path forward. We would refrain from being so presumptuous from making that decision."

If your county has not experienced a ransomware attack, consider yourself lucky. Here's a look at some recent attacks on county government systems:

Butler County, Kan. employees noticed an attack to their computer network after 911 operators and jail staffers saw errors pop up on their screens, County Administrator William Johnson said. "They absolutely shut down our network. We believe they struck on a Saturday because they knew there would be fewer people working."

"We were pretty embarrassed by this originally," he said. A data restoration company the county is using told them that these kinds of attacks are almost impossible to stop. "It's just how difficult you can make it for them." The attack crippled the county's motor vehicle, driver's license and register of deeds operations, as well as computerized warrants and inmate records, Johnson said.

In the future, Johnson said, the county will likely step up its employee training regarding suspicious-looking emails. "That is where we have failed," he said. "It's like a fire drill. It's a sad day in our society when we have to do something like this."

An attack on Schuyler County, N.Y.'s, computer system prompted an investigation by the FBI and the county hired a private cyber security law firm, Mullen Coughlin LLC of Wayne, Pa. Unlike Butler County, Kan., no county department was ever shut down, but some features of the county's 911 system were impacted, such as mapping.

In Becker County, Minn., thanks to the county's purchase last year of a backup and continuity system, the county was able to save data and allow IT personnel to quickly retrieve it after a ransomware attack in August. The $89,000 backup system allowed the county to restore its network within about 24 hours and get its website, printers and network back up and running in a few days. As soon as the county discovered the attack, they shut down their connection to law enforcement and other agencies, White said.

FBI Tips on Dealing with Ransomware

So, what does the FBI recommend you do to prevent a ransomware attack? As ransomware techniques and malware continue to evolve—and because it's difficult to detect a ransomware compromise before it's too late—counties should focus on:

- Awareness training for employees.
- Robust technical prevention controls.
- Creation of a solid business continuity plan in the event of a ransomware attack.

- Make sure employees are aware of ransomware and of their critical roles in protecting the organization's data.
- Patch operating system, software, and firmware on digital devices (which may be made easier through a centralized patch management system).
- Ensure antivirus and anti-malware solutions are set to automatically update and conduct regular scans.
- Manage the use of privileged accounts—no users should be assigned administrative access unless absolutely needed—and only use administrator accounts when necessary.
- Configure access controls, including file, directory, and network share permissions appropriately. If users only need read specific information, they don't need write-access to those files or directories.
- Disable macro scripts from office files transmitted over email.
- Implement software restriction policies or other controls to prevent programs from executing from common ransomware locations (e.g., temporary folders supporting popular internet browsers, compression-decompression programs).
- Back up data regularly and verify the integrity of those backups regularly.
- Secure your backups. Make sure they aren't connected to the computers and networks they are backing up.

21. The Two Faces of Social Media*

Martha Perego

The infiltration of social media into our personal lives and workplaces is creating some interesting ethical issues for local government organizations. Organizations have always had an overriding interest in and an expectation of controlling who speaks and what's said on behalf of the organization. Social media networks now make for quick communication to an exponentially large audience but often without that distinct line between personal opinion and official position. It's highly likely that you will encounter one of these situations.

Q: The city manager learned that a library employee started a Facebook page promoting the idea of a new library for the city. This news surfaced after a councilmember was asked to be a "fan" of the page. The manager checked the site and discovered that the employee was indeed listed as the administrator of the page. It was not clear from the information posted whether the employee had direction or approval of the library board to set up the site. The manager was uncomfortable with a city entity or employee promoting the development of a new facility without engagement with the city council, and he sought advice about how to proceed. Should he order the employee to take down the site?

A: Regardless of whether it is via a blog, Facebook, or the now seemingly archaic letter to the editor, all employees have a First Amendment right to comment, after hours and on their own equipment, about city matters. Even though this might create discomfort at the leadership level, the activity cannot be banned outright.

---

*Originally published as Martha Perego, "The Two Faces of Social Media," *Public Management (PM) Magazine*, June 2010 issue. Reprinted with permission of the publisher.

Probably a wise approach would be to talk with the employee about the importance of staff working in concert with the leadership on projects rather than running a solo effort, and that is what the manager in this case did. If the library staff person is really interested in helping the cause, she might be willing to take the site down and rethink her strategy.

Upon reflection and after a conversation with the manager, the employee decided it was best to turn the task of administering the site over to a member of the local "friends of the library," especially since the effort would have mandated that she stay on top of the site during work hours. She continues to be a proponent of a new library but on her own time.

Would the manager take a different tack if the employee in question was the library director? Certainly. Management personnel represent the organization, and when they speak, their comments should reflect the official position of the agency. It would be inappropriate for the library director to promote a concept that had not been vetted with the city manager and governing body.

Social media sites, when branded with city logos or authored by staff, are an extension of the local government's information network. For that reason, the organization has a vested interest in making sure that the information presented represents its official position.

Q: The human resources (HR) director received a packet of information from an anonymous source containing blog entries allegedly written by the assistant city manager. Using a catchy pseudonym, the author commented on news articles and other postings ranging from foreign policy to cultural issues of the day. None of the postings related to city matters.

As she read through the entries, the HR director was startled by the blunt, direct, and at least in her assessment, rude comments made by this individual. If these entries were indeed the work of the assistant city manager and his identity was revealed, it would not reflect well on the city. But since none of the entries was related to city matters, should the HR director pursue this?

A: The ICMA Committee on Professional Conduct had the opportunity recently to weigh in on a similar scenario. The committee noted that Tenet 7 of the ICMA Code of Ethics gives members the leeway to voice their opinions on policy matters or on issues of the day. If the policy matter relates to the ICMA member's workplace, the member ought to carefully consider the impact of expressing a personal opinion although the member is still free to do so.

With regard to blogging about nonemployer situations, the committee offered members this guidance:

- Use of a pseudonym is okay but be cognizant that anonymity is never guaranteed.

- It is not appropriate to use municipal networks to express personal opinions.
- Don't leverage your position or use your title.
- Think carefully about how comments or views may reflect on your ability to perform, your organization, your profession, and your professionalism.
- Consider the principles of Tenet 3 of the ICMA Code of Ethics, which encourage members to conduct themselves in all professional and personal matters in a manner that promotes public confidence in the profession.

Cities and counties are just now starting to develop policies to provide guidance on the official use of social media, and many readily acknowledge that what appears to be personal communication can bleed over into the work world. For that reason, some organizations ask employees to omit references to their place of employment or their job titles on personal sites.

Other communities are more liberal in allowing references to work and accomplishments but with the caveat that all content should be consistent with the organization's values and professional standards. They make it clear that if the writer does not follow those guidelines, it could affect future advancement and employment. Blog and twitter away … please just be civil and clear whether you are expressing a personal opinion or speaking on behalf of the organization you serve.

22. "Zero-Day" Stockpiling Puts Us All at Risk*

Benjamin Dean

"Zero-days" are serious vulnerabilities in software that are unknown to the software maker or user. They are so named because developers find out about the security vulnerability the day that it is exploited, therefore giving them "zero days" to fix it.

These vulnerabilities can be found in some of the most widely used software and platforms on the commercial market: Adobe Flash, Internet Explorer, social networks (Facebook and LinkedIn, to name two) and countless others.

The recent dump of emails from Hacking Team sheds new light on the extent of government involvement in the international market for zero-days. Rather than disclosing these vulnerabilities to software makers, so that they can be fixed, government agencies buy and then stockpile zero-days.

This practice and the policy that permits it expose billions of internet and software users to serious and unnecessary cybersecurity risks. A number of solutions to this problem are available, but first let's take a look at the zero-day market.

The Growing Market for Zero-Days

Knowledge of the existence of zero-days is valuable to criminals and intelligence agencies alike. They pay lots of money to learn about these vulnerabilities and then develop exploits (or simply purchase the exploits) to circumvent the information security of their targets.

---

*Originally published as Benjamin Dean, "'Zero-Day' Stockpiling Puts Us All at Risk," *The Conversation*, August 4, 2015. Reprinted with permission of the publisher.

Among other techniques, the hackers that breached Sony Pictures Entertainment and the Office of Personnel Management (OPM) exploited zero-day vulnerabilities to pull off these high-scale hacks.

This has become serious business. The international market for the buying and selling of zero-day vulnerabilities comprises three overlapping markets: "black," "gray" and "white."

Sellers in the black market include freelance hackers and organizations. Buyers include criminals and criminal organizations. Given the underground nature of the market, there's no telling how many vulnerabilities are bought and sold on the black market. Roy Lindelauf, a researcher at the Netherlands Defence Academy, believes that more than half of exploits sold are now bought from bona fide firms rather than from freelance hackers, suggesting that the black market is not the biggest of the three interlinked markets.

The second market is "gray" in the sense that it is legal though unofficial and unregulated. Nation-states historically have had a monopoly over buying in the gray market. They include Brazil, India, Israel, Malaysia, North Korea, Russia, Singapore, the United Kingdom, the United States and many more. Defense contractors such as Northrupp Grumann and Raytheon are also thought to be buyers and/or sellers.

Firm estimates of the size of the gray market are difficult to make. The National Security Agency (NSA) in the United States is considered to be "the best, surest zero-day acquirer ... in truth, a really insatiable one," according to a Hacking Team email indexed by WikiLeaks. It spent U.S.$25 million in 2013 to procure "software vulnerabilities" from private malware vendors. One source suggests that the average price for a zero-day ranges from $40,000 to $160,000.

Buyers in the also legal "white" market include software makers such as Facebook, Google, Microsoft and LinkedIn. Software makers offer a sum of money, sometimes called "bug bounties," to anyone who finds and discloses the existence of a vulnerability to them.

There are also platforms that connect dozens of software makers with security researchers and experts. They promise a commission to those who disclose vulnerabilities to software makers through the platform. iDefense and TippingPoint were two early companies in this space. New companies have joined the scene, such as HackerOne, which recently raised $25 million in venture capital.

Bug bounties are a novel solution to the problem of zero-days: pay people not to hack a system. Instead, pay those people to use their skills to find and disclose vulnerabilities so that software makers can fix them, thereby improving overall cybersecurity.

The amounts paid through bug bounty programs can be significant. In all markets, prices tend to be determined by the type of bug and the potential for hacking use. However, the prices on the white market are not typically as high as prices on the black market, nor do the prices come close to the losses incurred by the victims of zero-day exploits.

Risks of Government Stockpiling

While many government agencies are buyers in the global gray market for zero-days, almost no countries have an explicit policy stance toward what they do with the bugs that they buy.

In the U.S., some details of the official policy toward disclosure of zero-days have been made public. Former NSA Director General Keith Alexander has stated that the agency uses zero-days "for defense, rather than … for offensive purposes." President Barack Obama's view, according to his advisers, is that "when the National Security Agency discovers major flaws in internet security" it "should—in most circumstances—reveal them … rather than keep them mum so that the flaws can be used." A broad exception, however, is made for a clear national security or law enforcement need.

The use of the phrase "national security" is curious considering that a policy of withholding any zero-days at all effectively puts the security of all users of the software in question—which in today's world includes companies, government agencies and individuals—at additional risk of being hacked.

To its credit, the U.S. has gone further than all other governments in explaining its policy toward zero-day disclosure. Australia, China, Russia and the United Kingdom have not made their stance on zero-days public at all.

The consequences of this practice—and the often-murky policies that permit it—are severe. When knowledge of a zero-day is bought and then stockpiled by a government agency, there's no guarantee that another malevolent person or organization might not discover (or purchase) and exploit that same vulnerability.

By withholding knowledge of zero-days, government agencies keep all software users in a state of suspended risk. The scope of this risk is global, as the software and platforms in question are used by billions of people.

What Alternatives Are There?

Instead of a policy of stockpiling zero-days, and the risks that this policy entails, what alternative policies might exist?

Mandatory disclosure, or greater oversight, over the discovery or purchase of zero-days are obvious domestic alternatives to the status quo. At an international level, "voluntary collective action to harmonize export controls on zero-days through the Wassenaar Arrangement" is seen as another possible direction, particularly given that it is currently under review. This agreement was designed to control the export and import of weapons and technologies that have potential military applications.

Computer security analyst and risk management specialist Dan Geer has proposed that the U.S. government outbid (by 10 times) every other buyer in the international market for zero-days so long as bugs are "sparse not dense" (that is, the software in question has few, not many, bugs).

If the NSA spends $25 million a year on zero-days, under Geer's plan this would increase to at least $250 million. The NSA budget is at least $10 billion annually, with $1.2 billion spent in 2013 on offensive cyber-capabilities (in other words, state-sponsored hacking).

Given the size of these budgets, Geer's proposal is financially possible, though it would require a serious change of official policy, starting with mandating the immediate disclosure of all bugs to software makers so that they can be patched.

Going for the Root

If governments were really serious about addressing the problem of zero-day vulnerabilities, they might consider going to the root of the problem: placing liability on software makers for buggy code.

The common practice for software makers, since the 1980s, is known as "patch and pray." In short, software makers rush a product out the door, opting to release patches for vulnerabilities later, instead of investing time and resources for additional testing and patching of bugs (including zero-days) before release.

The economic logic is simple. Shipping equals sales and revenue. Delaying release to test and correct bugs adds to costs. Given that the losses from faulty software fall on the user, not the software maker, there's little incentive for the software maker to fix the bugs before shipping. It's easier to "move fast and break things" when you don't have to pay for the things that end up broken.

To make matters worse, users do not always promptly update their software, which is really the only defense they have. Vulnerabilities can thus persist for years after they have been discovered and patches made available.

Placing liability on the software maker for the losses due to their buggy software would completely alter these incentives. A number of approaches could be investigated in an attempt to find one that balances the need to minimize bugs, and protect users, while not smothering innovation.

Placing any kind of liability on software makers for their faulty products would take a great deal of political will, particularly in a climate where current proposals are pushing for the opposite. However, if done correctly, it would create a strong incentive for software makers to adopt more rigorous measures to reduce the number of bugs in their software. This would give a meaningful boost to the cybersecurity of billions of software users.

Paradox of Cybersecurity Policy Continues

Government officials claim to be doing everything possible to enhance cybersecurity. Zero-days are a serious threat to the cybersecurity of individuals, government agencies and corporations.

Yet government agencies are the biggest buyers of zero-days. If they're serious about cybersecurity, why then do these government agencies withhold knowledge of some of the zero-days that they discover or purchase?

This is yet another example of the paradox of current cybersecurity policy: government agencies tasked with enhancing cybersecurity conduct activities that result in the opposite outcome.

A clear policy of disclosure of all discovered or purchased zero-days would be a major step forward in bolstering cybersecurity internationally. Even better would be a policy that goes to the root of the problem, by allocating some liability on software makers for the losses linked to their buggy software.

Until the political will is mustered to address the problem of buggy software, including zero-days, the best that software users can do to protect themselves, unfortunately, is to follow the software makers' lead: patch and pray.

Part III. Prevention, Protection and Partnership

A. Practical Preventive Measures

23. A Plan for Cybersecurity*

Cory Fleming

At Issue

Computer systems represent critical infrastructure for all local governments. When a computer system goes down, it's difficult for staff to get work done. Recreating data stolen or lost in a cyber attack is time-consuming and labor intensive. With stories about cybersecurity attacks, security breaches, and information leaks popping up in the news on nearly a daily basis, local government managers have good reason to fear possible attacks to their computer systems and the havoc the attacks could cause.

All organizations—big and small—are vulnerable to attacks. Instituting cybersecurity programs and procedures is a bit like buying insurance from a risk management standpoint: You hope you never encounter attacks or breaches but you want to be prepared if you do. The challenge is to stay ahead of technology that continually morphs into new threats. At the same time, managers cannot be concerned only about the most recent attack or threat. They must look at the big picture to determine what it takes to always be operating in a secure environment.

Consideration needs to be given to how managers can respond to keep systems safe while not breaking their budgets; how local governments can-

---

*Originally published as Cory Fleming, "A Plan for Cybersecurity," Chapter 3, in *Cybersecurity: Protecting Local Government Digital Resources* (Washington, DC: ICMA and Microsoft, 2017). Reprinted with permission of the publisher.

retain qualified IT staff when private sector salaries are significantly higher; and for small community managers who also wear the hat of chief information officer (CIO), how they can stay abreast of trends and choose the most practical option.

Challenges

Workforce. Multiple studies have found that the lack of skilled IT personnel in the public sector is a significant problem. High salaries in the private sector make it difficult to attract IT talent to public sector positions and then also retain them, especially those individuals specializing in cybersecurity. Further aggravating the situation, leadership within the IT public sector is aging and many talented individuals have begun retiring in force. "A slow but persistent drain on human resources looms large, with many states reporting that 20 to 60 percent of their IT employees are nearing or at the age of retirement," according Derek Johnson, a state and local analyst with Deltek.[1]

An available workforce is simply not there for local governments to hire. "Thirty-five percent of organizations have open security positions that they are unable to fill and 53 percent say it can take as long as six months to fill one need, according to 'The State of Cybersecurity: Implications for 2015,' a study by ISACA," reports Marc van Zadelhoff of IBM Security.[2]

Budgeting. Cybersecurity is a critical issue for many communities that may not have the necessary funding or resources available to protect computer systems. While Noelle Knell reports in Government Technology that "[c]ities are expected to spend $30.9 billion on IT in 2017…and counties $22 billion,"[3] Paul Lipman, CEO of iSheriff, points out that "[t]he typical state or local government agency spends less than 5% of its IT budget on cybersecurity, compared to over 10% in the typical commercial enterprise."[4]

However, many security measures are simple and low-cost, such as following good password practices, changing system passwords on a routine basis, keeping browsers and operating systems updated, and using two-factor authentication systems (a passcode and a security question) where possible. In addition, training and education—helping staff understand what to watch for to prevent attacks—can help ward off new threats.

Plan Elements

Research and Needs Assessment. Research on current technology solutions is crucial for understanding the wide range of products available and how well those products can be expected to perform over time. Local governments should avoid quick fixes and one-off solutions. Instead, staff should research how products can be integrated with existing systems to produce better results.

Likewise, a needs assessment can help IT professionals determine what their requirements for technology solutions should be. A needs assessment begins with inventorying legacy systems currently being used and identifying where security gaps may exist. For example, is the credit card payment system protected from hackers trying to enter through the city's website? Local governments need to understand their cyber vulnerabilities and what it takes to mitigate those risks.

Policies and Procedures. Local governments need to develop written policies and procedures on how the organization will protect its computer systems. Buying software solutions or electronic devices alone will not protect a system. Employees must understand that human error plays a substantial role in many security breaches. Clicking on suspicious links in e-mails or not using a secure passcode to lock a computer screen when away from the computer exposes the organization's system to unneeded risk. While people are human and make mistakes, routine reminders of the organization's policies and procedures keep security measures at the forefront of staff concerns.

Roles and Responsibilities. Successful recovery from a cyber attack or other security incidents involving local government computer systems requires planning and preparation in the same way that recovery from natural and human-made disasters do. When employees know, before an incident occurs, who is responsible for what tasks and what actions will be top priorities, the response goes more smoothly and quickly. Establishing good working relationships before an event is helpful when determining what resources are available and which staff have needed skill sets to respond.

Training and Education Programs. As noted earlier, the number of openings in the cybersecurity field far outpace the number of trained professionals available to fill those positions. This general lack of IT talent makes it imperative for local governments to invest in training and education for existing staff. The level of training needed will vary among individuals, but team members need to be aware of system vulnerabilities, how and where they should look for potential risks, and the tools available to protect systems. At a minimum, all employees need to know and follow recommended safety standards.

System Integration. Computer systems, devices, and other technology solutions need to work together to achieve maximum protection. Often, however, staff purchase a one-off solution designed to solve a specific issue, but the solution won't work with related issues. Consider the relationship between an app for identity protection and one for theft protection. Both solutions provide a certain level of protection, but if the apps are integrated, the level of protection is far greater. "Cybersecurity needs to be integrated throughout an institution as part of enterprise-wide governance processes, information security, business continuity, and third-party risk management," according to the Federal Financial Institutions Examination Council.[5]

Security Audits. All local governments need to conduct regular security audits to determine what data may be at risk and enable staff to understand where threats and vulnerabilities exist in the system. Internal controls for IT are often overlooked, including, for example, limiting the number of staff who have access to business operations data, establishing security and privacy policies for staff, or maintaining password-protected locations within a system. Adherence to simple security practices like these ensures data integrity and protects the organization's systems by limiting opportunities for breaches. By identifying what data assets an organization maintains and classifying the importance of that data, IT staff can act to protect against those risks. Security audits also provide a good starting point for conducting a needs assessment to support the development of a cybersecurity plan. A needs assessment defines the tasks and activities that should take place to move a local government from "ready for the status quo" to "ready for the unknown."

Monitoring. Local governments are capturing and maintaining increasing amounts of data every year due to sensors, the Internet of Things (IoT), body-worn cameras, and other new technologies. The new data make it possible for local government professionals to analyze data and make better management and policy decisions based on data. Increased data sharing within and outside a local government organization represents new opportunities to adopt smart community management practices that range from developing smarter deployment of staff for routing lawn and turf management to the use of drones to collect high-resolution images for marketing business and industrial sites for development.

New data and sharing of that data, however, also open new opportunities to transmit viruses and could make a local government vulnerable to other malicious activity. Monitoring is used for threat detection, but risk specialists at Deloitte note that monitoring should be used to "proactively identify those activities most detrimental to the business and support mitigation decisions."[6]

Continuity of Operations. Public safety officials have routinely developed plans for the continuity of operations should a disaster hit their community. Just as leaders map out evacuation routes to prepare for flooding or where to set up shelters during a hurricane, so too must they consider which data and computer systems are most critical for daily operations or how they can rebuild systems as quickly as possible in the event of a cyber attack.

Backup plans for what to do in the event of a digital attack are as important for keeping a community operational as are those for roads and shelters during a storm.

Communications. Communication is critical when any emergency or disaster hits. Local government managers need to be prepared to communicate using different channels during a natural disaster. During a cyber attack, the need to have access to multiple forms of communication is even more relevant. Email, texts, instant messages, and other electronic channels may be shut down during an incident. In addition to being physically able to communicate and coordinate within the organization, local governments also need to be prepared to brief elected officials, the news media, and the public about what happened, how it was contained, and what, if any, damage was sustained. A clear, concise, and accountable message after an event will go a long way to reassuring stakeholders that their personal data are protected.

Responding to a Breach

Setting Priorities for Systems Restoration. Each community will have different priorities for systems restoration. For almost all communities, however, systems delivering critical citizen services—such as health (a county hospital) or safety (police communications)—will be first among identified priorities. Police, fire, and emergency medical personnel need to be able to communicate and access their data quickly.

Backup Data. Most local government organizations routinely back up data in their systems, but often employees don't back up their local drives to prevent loss of programs in the event of an attack. Field crews using mobile technology such as smartphones and tablets should be routinely backed up, saving both data as well as program files. Employees who use laptops should also perform full backups on their machines daily.

Sharing Resources. Unlike the private sector, which must contend with business competition, local governments and other public agencies can and should work together. Whether it is sharing leading practices or splitting costs to purchase necessary software, local governments have options available to them that businesses don't.

Assessment and understanding of your organization's levels of cybersecurity take the same kind of planning that you invest in budgeting and strategic planning. The old adage—"By failing to prepare, you are preparing to fail"—is true. Strategic planning enables local government leaders and staff to prepare for a more secure future.

Notes

1. Derek Johnson, "State, Local Governments Turn Attention to Cybersecurity Capabilities," *The Washington Post,* April 26, 2014. https://www.washingtonpost.com/business/capitalbusiness/state-local-governments-turn-attention-to-cybersecurity-capabilities/2014/04/04/8527c4b0-b912-11e3–899e bb708e3539dd_story.html?utm_term=.b1dca4b314de

2. Marc van Zadelhoff, "Four Big Cyber Security Challenges (and How to Overcome Them)," *Forbes*, May 14, 2015. https://www.forbes. com/sites/ibm/2015/05/14/four-big-cyber-security-challenges-and-how-to-overcome-them/#5771430e5867

3. Noelle Knell, "IT Spending in State and Local Government: What Does 2017 Hold?," *Government Technology*, March 20, 2017. http://www.govtech.com/budget-finance/IT-Spending-in-State-and-Local-IT-What-Does-2017-Hold.html

4. Lipman, Paul, "The Cybersecurity Challenges Facing State and Local Governments," *infosecurity*, August 19, 2015. https://www.infosecurity-magazine.com/opinions/cybersecurity-challenges-state/

5. Federal Financial Institutions Examination Council, FFIEC Cybersecurity Assessment Tool, May 2017. https://www.ffiec.gov/pdf/cybersecurity/FFIEC_CAT_May_2017.pdf

6. Christopher Stevenson, Andrew Douglas, Mark Nicholson, and Adnan Amjad, "From Security Monitoring to Cyber Risk Monitoring: Enabling Business-aligned Cybersecurity," *Deloitte Review*, Issue 19, July 25, 2016. https://dupress.deloitte.com/dup-us-en/deloitte-review/issue-19/future-of-cybersecurity-operations-management.html

24. How We Can Each Fight Cybercrime with Smarter Habits*

ARUN VISHWANATH

Hackers gain access to computers and networks by exploiting the weaknesses in our cyber behaviors. Many attacks use simple phishing schemes—the hacker sends an email that appears to come from a trusted source, encouraging the recipient to click a seemingly innocuous hyperlink or attachment. Clicking will launch malware and open backdoors that can be used for nefarious actions: accessing a company's network or serving as a virtual zombie for launching attacks on other computers and servers.

No one is safe from such attacks. Not companies at the forefront of technology such as Apple and Yahoo whose security flaws were recently exploited. Not even sophisticated national networks are home free; for instance, Israel's was compromised using a phishing attack where an email purportedly from Shin Bet, Israel's internal security service, with a phony PDF attachment, gave hackers remote access to its defense network.

To figure out why we fall for hackers' tricks, I use them myself to see which kinds of attacks are successful and with whom. In my research, I simulate real attacks by sending different types of suspicious emails, friend-requests on social media, and links to spoofed websites to research subjects. Then I use a variety of direct, cognitive and psychological measures as well as unobtrusive behavioral measures to understand why individuals fall victim to such attacks.

*Originally published as Arun Vishwanath, "How We Can Each Fight Cybercrime with Smarter Habits," *The Conversation*, January 26, 2015. Reprinted with permission of the publisher.

What is apparent over the many simulations is how seemingly simple attacks, crafted with minimal sophistication, achieve a staggering victimization rate. As a case in point, merely incorporating the university's logo and some brand markers to a phishing email resulted in close to 70% of the research subjects falling prey to the attack. Ultimately, the goal of my research is to figure out how best to teach the public to ward off these kinds of cyberattacks when they come up in their everyday lives.

Clicking Without Thinking

Many of us fall for such deception because we misunderstand the risks of online actions. I call these our cyber-risk beliefs; and more often than not, I've found people's risk beliefs are inaccurate. For instance, individuals mistakenly equate their inability to manipulate a PDF document with its inherent security, and quickly open such attachments. Similar flawed beliefs lead individuals to cavalierly open webpages and attachments on their mobile devices or on certain operating systems.

Compounding such beliefs are people's email and social media habits. Habits are the brain's way of automating repeatedly enacted, predictable behaviors. Over time, frequently checking email, social media feeds and messages becomes a routine. People grow unaware of when—and at times why—they perform these actions. Consequently, when in the groove, people click links or open attachments without much forethought. In fact, I've found certain Facebook habits—such as repeatedly checking newsfeeds, frequently posting status updates, along with maintaining a large Facebook friend network—to be the biggest predictor of whether they would accept a friend-request from a stranger and whether they would reveal personal information to that stranger.

Such habitual reactions are further catalyzed by the smartphones and tablets that most of us use. These devices foster quick and reactive responses to messages though widgets, apps and push notifications. Not only do smartphone screen sizes and compressed app layouts reduce the amount of detailed information visible, but many of us also use such devices while on the go, when our distraction further compromises our ability to detect deceptive emails.

These automated cyber routines and reactive responses are, in my opinion, the reasons why the current approach of training people to be vigilant about suspicious emails remains largely ineffective. Changing people's media habits is the key to reducing the success of cyberattacks—and therein also lies an opportunity for all of us to help.

Harnessing Habits to Fight Cybercrime

Emerging research suggests that the best way to correct a habit is to replace it with another, what writer Charles Duhigg calls a Keystone Habit. This is a simple positive action that could replace an existing pattern. For instance, people who wish to lose weight are instructed to exercise, reduce sugar intake, read food labels and count calories. Doing this many challenging things consistently is daunting and often people are too intimidated to even begin. Many people find greater success when they instead focus on one key attainable action, such as walking half a mile each day. Repeatedly accomplishing this simple goal feels good, builds confidence and encourages more cognitive assessments—processes that quickly snowball into massive change.

We could apply the same principle to improve cybersecurity by making it a keystone habit to report suspicious emails. After all, many people receive such emails. Some inadvertently fall for them, while many who are suspicious don't. Clearly, if more of us were reporting our suspicions, many more breaches could be discovered and neutralized before they spread. We could transform the urge to click on something suspicious into a new habit: reporting the dubious email.

We need a centralized, national clearing house—perhaps an email address or phone number similar to the 911 emergency system—where anyone suspicious of a cyberthreat can quickly and effortlessly report it. This information could be collated regionally and tracked centrally, in the same way the Department of Health tracks public health and disease outbreaks.

Of course, we also need to make reporting suspicious cyber breaches gratifying, so people feel vested and receive something in return. Rather than simply collect emails, as is presently done by the many different institutions combating cyber threats, submissions could be vetted by a centralized cybersecurity team, who in addition to redressing the threat, would publicize how a person's reporting helped thwart an attack. Reporting a cyber intrusion could become easy, fun, something we can all do. And more importantly, the mere act of habitually reporting our suspicions could in time lead to more cybersecurity consciousness among all of us.

25. Simple Steps to Online Safety*

ALAN SHARK

October has been designated Cyber Security Awareness Month and the U.S. Department of Homeland Security has issued a series of 5 weekly topical themes. This week's theme in Simple Steps to Online Safety NACo in partnership with the Public Technology Institute (PTI) have developed a series of useful checklists and commentary that are specifically aimed at the public manager.

Cyber security breaches have grown some 26 percent over last year with ransomware continuing to rise. Local governments have always been particularly attractive targets since they collect and store such massive amounts of personal information. And with more mobile devices and social media apps, there are more entry points for mischief than ever before.

According to research from Egress Software Technologies most data breach incidents in local government were caused by human error. There some rather simple and straightforward steps one can take to protect themselves as cyber security awareness most always starts with the individual. It is important to keep in mind there is another group that is observing us—our employees. We must set an example by following key best practices ourselves. Surely we can't expect our employees to adhere to best practices for mitigating cyber threats when we ourselves exempt ourselves. Here we must lead by example.

- **Passwords still matter**. Using different passwords that contain and include at least 8 characters with numbers and symbols. Try and

*Originally published as Alan Shark, "Simple Steps to Online Safety," *NACo County News*, Oct. 3 & 16, 2017. Reprinted with permission of the publisher.

come with a formula where you can remember them too. For example, you may use and old address as a starter or transpose a letter for a number or symbol. Passwords should not begin with a capital letter, and underscore is a good way to separate a bunch of numbers.

- **Use Multiple Passwords.** By using multiple passwords for different accounts, you spread the risk of having one breach expose you to everywhere you have a login account. Too many passwords to remember? Consider using a "password manager" like LastPass or Dashlane. Most offer free versions that one can try out. While these systems require a complex master password, password managers do the rest. You can elect to have them assign complex random passwords and most have an autofill feature that fills in the necessary fields automatically. Another advantage is most password managers remember and recall passwords and payment information across your devices if you so choose. This includes PC, laptops, and all your mobile devices.
- **Think Before You Click.** Ransomware and phishing attacks have increased dramatically the past two years. Many of these attacks can be traced to employees clicking and opening attachments. Before you open an attachment are you sure it is from a person or entity they say they are? Do you see suspicious signs like misspellings, using a salutation such as "dear customer" instead of your name, a return URL/address that is different from the senders? For example, if you receive something that appears to be from your bank, is the URL taking you to the bank or is it directing you somewhere else. It's always best not to click on such emails regardless of how real they look. Instead simply go directly to the company's site and see if there is any real issue for you to resolve. Finally, if in doubt always contact your IT folks as they have ways of checking authenticity without risk to others.
- **Limit Address Book Entries.** It is shocking to learn how many professionals use their mobile device address books to store credit card numbers, passwords, family social security numbers and birthdates. As temping as it is don't use your mobile device's directory as your personal information database! Most cyber breaches attack your address books and yes, these same rogue software programs are programmed to search for this type of information in addition to all your contacts. Remember, the badguys goal is to exploit ever bit of information they can and use it to cause further havoc which could lead to identity theft, use passwords to enter systems to obtain further a perhaps more important information.

- **Update Your Devices.** Computer and mobile device manufacturers are routinely updating their operating systems to help improve performance as well as actively addressing known security vulnerabilities. It should go without saying, make sure you not only have the best virus and malware protection—but it is updates in real-time to gain maximum protection.
- **Avoid Public WI-FI.** It is always tempting for on-the-go-people to connect every time they see a Wi-Fi hotspot. There are plentiful offerings at airports, trains, coffee shops, hotels, and conferences. Unfortunately, public Wi-Fi (free or not) can easily be exploited by the bad-guys who can "see" what you are logging into with not much effort and be able grab your passwords. Never conduct business in public places offering Wi-Fi that requires passwords which might include logging into your office or your bank. Consider having your own mobile hotspot offered by all wireless carriers. Even though you are still connecting via Wi-Fi it is far more difficult to snoop and the data is usually encrypted and ultimately converted to more secure cellphone frequencies.

Cyber Security in the Workplace Is Everyone's Business

October has been designated Cyber Security Awareness month and the U.S. Department of Homeland Security has issued a series of five weekly topical themes. This week's theme is "Cyber Security in the Workplace is Everyone's Business." NACo, in partnership with the Public Technology Institute (PTI) has developed a series of useful checklists and commentary created for county elected leaders.

On Week 1, published in the Oct. 3 issue of *CN Now*, we focused on what an individual can do to be more cyber secure. This week we will focus on what an organization can and must do.

Cyber security breaches have grown some 26 percent over last year with ransomware continuing to rise. County governments have always been particularly attractive targets because they collect and store such massive amounts of personal information (tax records and payment information, for example). With the growth in the use of mobile devices and social media apps, there are now more entry points for mischief than ever before.

The weakest link continues to be our employees. One misguided click on a targeted phishing email can compromise an entire organization. To make matters worse, many phishing emails tend to come from employees whose names we know and whose email address has become compromised in an earlier attack.

Recommendations that affect individuals are largely the same, however, with an added emphasis of the potential impact on an entire organization. One careless staff person can bring down an entire county operation.

Many counties require cyber security awareness training while others simply provide optional training. Our experience shows that many programs are inadequate for several reasons, which include:

- Training is only required once a year.
- Training can be too technical.
- Training can scare some staff and can create an environment of resentment or fear of punishment.
- Training can lack real-world examples and is often out-of-date.

While much of the actual protection of the digital infrastructure resides with the technical experts, there are two paramount roles county elected leaders can and should play. The first one is for public officials to set the proper example themselves. This means following the rules such as having and changing complex passwords.

The second role is to ensure a safe and secure cyber environment. The key component of this is to have a robust cyber security awareness program. Many programs offered today online or in person vary in quality and approach. Many public officials ask, what should I be looking for and what are the elements of a sound cyber security awareness program? Here is a list to consider.

Assign a senior staff member to be in charge. This person might be the chief information officer, the chief information security officer, or other designee who is both technical and people-oriented. A high-level administrator or HR professional can also fill this role.

The best plans are ongoing and not just an annual event of a few hours of training.

Practice the elements of the plan and conduct drills to make sure everyone understands and complies.

Make sure there are stated consequences for careless behavior, depending on the levels of any violation.

While making sure you hold to your stated policies and procedures, you also want to make sure that you create a positive environment that encourages staff to report things at once if they believe they may have come across something wrong. In fact, there should be consequences for anyone not reporting an incident immediately.

Conduct regular, focused sessions aimed at exploring various types of cyberattacks. This will help demonstrate your organization's commitment to keeping systems safe as well as to keep the topic front and center with employees.

Consider role playing to help demonstrate how criminal elements use the phone, or social media to manipulate staff into providing valuable data that get into the wrong hands.

Employees should be trained to recognize an attack; to know not only what it looks like, but who to call and when to report the attack.

Always encourage employees to come forward with anything that they feel does not look or feel right. There have been many cases where an alert employee reported something as it was unfolding and as a result was able to minimize damage and loss.

Overall, training must be relevant and should be fun—like playing detective or guarding the "palace" as in a video game.

There are many digital destinations one can turn to for more information and assistance.

Check this story online at www.countynews.org for some very useful resources. Some are a bit more technical—so if you think it is useful, simply pass it on to your technical staff—it shows your interest. Remember Cyber Security Awareness is about awareness.

Finally, make sure your organization is a member of MS-ISAC, a NACo and PTI partner; membership is free and they are funded by DHS.

26. Staying Safe on Social Networking Sites*

U.S. Computer Emergency Readiness Team

What Are Social Networking Sites?

Social networking sites, sometimes referred to as "friend-of-a-friend" sites, build upon the concept of traditional social networks where you are connected to new people through people you already know. The purpose of some networking sites may be purely social, allowing users to establish friendships or romantic relationships, while others may focus on establishing business connections.

Although the features of social networking sites differ, they all allow you to provide information about yourself and offer some type of communication mechanism (forums, chat rooms, email, instant messenger) that enables you to connect with other users. On some sites, you can browse for people based on certain criteria, while other sites require that you be "introduced" to new people through a connection you share. Many of the sites have communities or subgroups that may be based on a particular interest.

What Security Implications Do These Sites Present?

Social networking sites rely on connections and communication, so they encourage you to provide a certain amount of personal information. When

*Public document originally published as U.S. Computer Emergency Readiness Team, "Staying Safe on Social Networking Sites," https://www.us-cert.gov/ncas/tips/ST06-003 (June 5, 2015).

deciding how much information to reveal, people may not exercise the same amount of caution as they would when meeting someone in person because

- the Internet provides a sense of anonymity
- the lack of physical interaction provides a false sense of security
- they tailor the information for their friends to read, forgetting that others may see it
- they want to offer insights to impress potential friends or associates

While the majority of people using these sites do not pose a threat, malicious people may be drawn to them because of the accessibility and amount of personal information that's available. The more information malicious people have about you, the easier it is for them to take advantage of you. Predators may form relationships online and then convince unsuspecting individuals to meet them in person. That could lead to a dangerous situation. The personal information can also be used to conduct a social engineering attack. Using information that you provide about your location, hobbies, interests, and friends, a malicious person could impersonate a trusted friend or convince you that they have the authority to access other personal or financial data.

Additionally, because of the popularity of these sites, attackers may use them to distribute malicious code. Sites that offer applications developed by third parties are particularly susceptible. Attackers may be able to create customized applications that appear to be innocent while infecting your computer or sharing your information without your knowledge.

How Can You Protect Yourself?

- Limit the amount of personal information you post—Do not post information that would make you vulnerable, such as your address or information about your schedule or routine. If your connections post information about you, make sure the combined information is not more than you would be comfortable with strangers knowing. Also be considerate when posting information, including photos, about your connections.
- Remember that the Internet is a public resource—Only post information you are comfortable with anyone seeing. This includes information and photos in your profile and in blogs and other forums. Also, once you post information online, you can't retract it. Even if you remove the information from a site, saved or cached versions may still exist on other people's machines.
- Be wary of strangers—The Internet makes it easy for people to

misrepresent their identities and motives. Consider limiting the people who are allowed to contact you on these sites. If you interact with people you do not know, be cautious about the amount of information you reveal or agreeing to meet them in person.

- Be skeptical—Don't believe everything you read online. People may post false or misleading information about various topics, including their own identities. This is not necessarily done with malicious intent; it could be unintentional, an exaggeration, or a joke. Take appropriate precautions, though, and try to verify the authenticity of any information before taking any action.
- Evaluate your settings—Take advantage of a site's privacy settings. The default settings for some sites may allow anyone to see your profile, but you can customize your settings to restrict access to only certain people. There is still a risk that private information could be exposed despite these restrictions, so don't post anything that you wouldn't want the public to see. Sites may change their options periodically, so review your security and privacy settings regularly to make sure that your choices are still appropriate.
- Be wary of third-party applications—Third-party applications may provide entertainment or functionality, but use caution when deciding which applications to enable. Avoid applications that seem suspicious, and modify your settings to limit the amount of information the applications can access.
- Use strong passwords—Protect your account with passwords that cannot easily be guessed. If your password is compromised, someone else may be able to access your account and pretend to be you.
- Check privacy policies—Some sites may share information such as email addresses or user preferences with other companies. This may lead to an increase in spam. Also, try to locate the policy for handling referrals to make sure that you do not unintentionally sign your friends up for spam. Some sites will continue to send email messages to anyone you refer until they join.
- Keep software, particularly your web browser, up to date—Install software updates so that attackers cannot take advantage of known problems or vulnerabilities. Many operating systems offer automatic updates. If this option is available, you should enable it.
- Use and maintain anti-virus software—Anti-virus software helps protect your computer against known viruses, so you may be able to detect and remove the virus before it can do any damage. Because attackers are continually writing new viruses, it is important to keep your definitions up to date.

Children are especially susceptible to the threats that social networking sites present. Although many of these sites have age restrictions, children may misrepresent their ages so that they can join. By teaching children about Internet safety, being aware of their online habits, and guiding them to appropriate sites, parents can make sure that the children become safe and responsible users.

27. Seven Keys to Strengthen Your Cybersecurity Culture*

DANIEL J. LOHRMANN

While running on my treadmill on Thursday morning, August 17, 2017, I was watching CNBC's *Squawk Box*, as David Novak, co-founder and former CEO of YUM Brands, came on the show as a guest.

He was asked how he was so successful at growing such a powerful set of global YUM Brands with great results including names like Pizza Hut, Taco Bell, Burger King and others. His answer made me slow the treadmill to a walk and listen closely.

He said several things, but his clear messages focused on building a great culture with a set of core values and staff recognition. Here's what stood out to me (paraphrased):

Success is all about the culture. Great leaders know your core values and are true to them. What messages are you sending to your employees? Are you recognizing and rewarding your staff?

As an aside: David Novak elaborates further on the recognition theme in this earlier article and video from last year. He challenged all of us to say "thank you" to employees and everyone in our lives more often. He even wrote this fun book on the 10 principles of recognition called O Great One!: A Little Story About the Awesome Power of Recognition.

Near the end of his *Squawk Box* interview, the topic of what actions to take on several global cybersecurity issues came up. Becky Quick asked Novak what the Trump administration should do about China stealing our intellectual property via computer hacking.

---

*Originally published as Daniel J. Lohrmann, "Seven Keys to Strengthen Your Cybersecurity Culture," *Government Technology*, August 20, 2017. Reprinted with permission of the publisher.

Novak said we need win-win answers that will work for both countries. Despite serious problems that require tough negotiations, we need to be positive in our approach, while enforcing laws and acting on areas where we have international agreement.

Issue One: Back to Security Culture

Management guru Peter Drucker is attributed with the well-known saying, "Culture eats strategy for breakfast." And while there are hundreds of books and thousands of articles on building great work cultures, not nearly as much is written about creating a positive enterprise culture emphasizing cybersecurity in the workplace.

So how can we lead a digital transformation that is also people-focused and security-focused at the same time? Here are a few of the common answers I have seen around the Internet over the past few years:

- Tripwire: 3 Tips on How to Create a Cyber Security Culture at Work
- Huffington Post: 6 Tips to Build a Cyber-Security Culture at Work
- Security Intelligence: Building a Cybersecurity Culture Around Layer 8

For several years now, the typical answers included a central focus on effective security awareness training for all employees as well as the need for management buy-in and business leadership for cybersecurity.

Nevertheless, digging a bit deeper, here are, in my view, seven keys to building a lasting security culture that can outlive individual security incidents and staff turnover.

1. **Genuine Executive Priority and Support**—We all know that children watch (and usually follow) what their parents do and not just what they say. In the same way staff learn what the real priorities are from executive actions. Are managers walking the talk? Are resources backing up the executive memos?

   For example, when I was CSO is Michigan government, Gov. Rick Snyder was a true champion for cybersecurity in the state, and in the nation, who frequently discussed cyberactions at cabinet meetings and led by example. If this executive priority focus is missing, you will struggle to succeed in the other areas in the long run. Consider these suggestions to build management support for cybersecurity.

2. **Honest Risk Assessment to Measure Security Culture Now**—What is the security posture currently? How are security audit findings addressed? What are real technology and security priorities? Are there metrics and/or dashboards to measure progress?

Also, this excellent article from Deloitte shows how to assess your culture from a perspective of beliefs, behaviors and outcomes.

3. A Clear Vision of Where You Want Your Security Culture to Be—A lot has been written about benchmarking and following best practices in cybersecurity. One important question is whether you know where you are heading. What is the vision of what success looks like for your security and technology teams?

Consider visiting your industry peers and learning from other public- and private-sector organizations that are doing cybersecurity culture well. Look at the National Association of State Chief Information Officers (NASCIO) award-winners, NGA best practices and state and local partners in your region. Consider a road trip to learn from others and benchmarking progress.

For example, back in 2011–2012, Stu Davis the Ohio CIO, brought a team up to Michigan to see how we built our security architectures and governance. Ohio state government used that visit and follow-on conversations to build an excellent cybersecurity program.

4. Do You Have a Cyber Plan?—Many state governments have published cybersecurity plans to clearly describe where they are going, who's involved, and what the expectations are for various groups. Examples include Michigan, Delaware, Missouri, North Carolina and others.

More details will soon be provided on this cybersecurity planning topic in an upcoming blog.

5. Clear Cybercommunication to the Masses—Great, you have a plan and specific actions steps. But does anyone know what's happening? What is the elevator pitch? How well are these messages received? Is the communication flowing both ways? Are you getting feedback?

Communicating cybermessages is an ongoing challenge, and no leader has done that better over the past year than Virginia Gov. Terry McAuliffe—who has made cybersecurity the top topic during his year as NGA leader.

6. End User Security Awareness Training for Everyone. This Includes Managers, System Admins and Other Specific Roles—As mentioned several times above, culture change definitely involves offering intriguing, relevant, updated, timely training that is brief, frequent and focused to the entire enterprise.

And while this is the area that is the one most often discussed regarding security culture change, it is only one component. Still, this cannot be a check-the-box exercise and be successful. I described this effective cybertraining area in much more detail in this recent interview with MicroAgility CEO Sajid Khan.

7. Celebrate Success with Food and Fun. Find out if security is a part of business DNA? How do you know what people are engaged in? Answer: See what they celebrate. When are their food and family showing up for awards?

Ask this question of your organization: When do you celebrate success? Assuming this is happening at all, are people rewarded for doing the right things regarding security? Any bonuses for great cyberetiquette or awards for doing the right things?

Here are some specific examples to ponder. And here are some cultural mistakes to avoid with security training.

Final Thoughts

In conclusion, building a healthy security culture is not a one-time project or one-year focus. Like building a great college football program at schools like Alabama, this is an ongoing challenge that must be repeated as the organization changes.

For more details, I really like this ISSA series of CISO mentoring talks, which provide many practical tips for security leaders to consider from CISOs who have been successful in different industries over many years. Following their advice is a great way to enhance your culture of cybersecurity.

Finally, I want to close with this quote from David Novak on the greatest challenge facing leaders today.

"Seven in 10 employees in the U.S. are not engaged. They're going to work and they can't wait to go home," he said.

Novak said great companies create environments where everyone counts and is valued.

That's why your corporate or government culture is so central to organizational success.

Is security a piece of your culture change efforts?

28. Three Tips for Forming a Computer Emergency Response Team*

KELSEY BREWER

Integration of technology into local government functions has increased at an exponential rate over the past 10 years. As cities continue to integrate smart technology into the day-to-day functions of local government, the need for strong safeguards against attacks are critical to maintaining system integrity.

Much of the conversation around cybersecurity for local governments focuses on avoiding attacks in the first place. Recommendations such as those discussed in the December 2017 issue of *LGR: Local Government Review* (powered by TownCloud) often include conducting threat assessments, investing in continuous monitoring systems, and training employees to avoid social engineering tactics often used by hackers to manipulate their way into sensitive systems. These are, of course, all valuable investments and steps to take when considering how to protect sensitive information from cyber-attack.

But what happens when, despite all the best planning and precautions, a cyber-attack is successful?

Almost every city in the United States has a Community Emergency Response Team. These programs help educate volunteers in disaster response skills such as fire safety, search and rescue, and basic first aid and were created to respond to physical and natural disasters in which, despite the best prepa-

*Originally published as Kelsey Brewer, "Three Tips for Forming a Computer Emergency Response Team," ICMA Blog, International City/County Management Association (January 17, 2018). Reprinted with permission of the publisher.

ration, an emergency situation requires a quick and efficient response. As cyber-attacks become more sophisticated, the likelihood of experiencing an emergency-level attack increases. Just as cities should invest in emergency responses to physical disasters, so should they invest in emergency responses to the digital world.

The concept of **Computer Emergency Response Teams** for local governments is presented by Microsoft in its *Developing a City Strategy for Cybersecurity: A Seven-Step Guide for Local Governments* whitepaper. These teams are comprised of experts from the private, government, and academic worlds and are meant to help coordinate the public response in case of a cyber incident. Computer ERTs take investments that cities have made on the front-end of cybersecurity (threat assessments, continuous monitoring, etc.) and use the information gathered from them to help the public and private sector respond in case of an emergency.

Threat assessments conducted by cities will show where there are vulnerabilities within systems, what information attackers are most likely to go after, and what should constitute as a high-level priority and low-level priority in case of an attack.

Computer ERTs then take this information to help assess credible threats and, in the case of an attack, take the lead on directing the cities' response. In California, the city of Los Angeles has embodied the idea of a Computer ERT in their Cyber Intrusion Command Center. Members meet on a regular basis to discuss common threats and possible methods of defusing them. But even smaller entities can employ the same method. Orange County, California's OC Intelligence Assessment Center has a dedicated team whose role is to not only assess cyber threats but to serve as the lead in post-incident mitigation, coordinating the necessary responses from both the public and the private sector.

As cyber-attacks become more common, the general public's expectation that their government is ready to react will only increase. In the past, claiming inexperience and unfamiliarity may have been a sufficient explanation for a lackluster emergency response to a cyber-attack. But as cities and their residents continue to integrate more and more technology into everyday functions, the need to plan for the inevitable becomes imperative. Anything short of enthusiastic preparedness becomes negligence of the digital disasters lurking underneath our keyboards.

Three Tips for Forming a Computer Emergency Response Team

1. Be Thoughtful about the Membership of Your Computer ERT

There are a couple things to consider when putting together the membership of your Computer ERT. The first is to ensure that all critical services that your city offers or contracts are represented on your team. Having representation from these groups will ensure a more robust dialogue when discussing how to protect these services from attack and how to bring them back online quickly should they be compromised. Cities should also consider geographical issues when developing a Computer ERT. Are there critical resources that you share with another city through a joint powers authority? Is the electrical grid control center that your city is dependent on located in another jurisdiction? If so, these outside entities should always be a part of your Computer ERT membership.

2. Create External Incident Classifications

Have your Computer ERT create incident classifications to ensure a quick and appropriate flow of resources. These classifications will determine when it is appropriate to notify the city of an incident or when it is important for the city to provide resources to external entities as part of the general response to an attack. While some Computer ERTs may focus on when an attack compromises functions or assets controlled by a local government, it is possible that an attack could target non-government-controlled resources that will impact a municipalities' ability to function. Some cities, for example, do not operate their own water district, but a threat to the local water transportation networks would certainly impact the ability of a city to operate. Classifying such an attack as "response-worthy" for a Computer ERT will help streamline resources if necessary.

3. Run Test Drills

Just like disaster preparedness, running test drills of response systems is critical to evaluating the effectiveness of current cyber-attack responses. Test drills should occur on a periodic basis and should evaluate the Computer ERT's response and ability to quickly coordinate internally and externally quickly and effectively.

29. Cybersecurity Strategies Governments Need*

Gus "Ira" Hunt *and* Lalit Ahluwalia

If there were any doubts about the critical need for governments, businesses and individuals to better fortify themselves against cyber threats, Petya should have put them to rest. The attack a few weeks ago using ransomware known by that name wreaked global havoc, infecting computers and networks in more than 65 countries including the United States.

The Petya outbreak followed—by just a few weeks—the even more widespread WannaCry ransomware attack. As evidenced by these high-profile events, protecting sensitive data and leveraging the right systems to detect, prevent and remediate security breaches continue to be a challenge for many organizations.

The concern is especially high for government agencies. As guardians of some of our most sensitive citizen and public-employee data, they are attractive targets for cyberattacks. Governmental organizations face dozens of focused, targeted attacks each year, one in three of which result in a successful security breach, according to a recent Accenture survey of security executives.

To bolster protection of our assets, government agencies must adopt modern, proactive, agile strategies that can help them quickly identify and respond to digital security risks. It's not clear, however, to what extent they are currently applying the right resources to confront this challenge.

A recent Accenture report based on a survey of 150 government executives in the United States suggests that most agencies don't have adequate technologies in place. Only 13 percent of respondents believe their existing

---

*Originally published as Gus "Ira" Hunt and Lalit Ahluwalia, "Cybersecurity Strategies Governments Need," *Governing*, August 3, 2017. Reprinted with permission of the publisher.

technology is effective for responding to cybersecurity breaches, and only one-third say they are confident in their ability to monitor, identify and measure these breaches. Almost half of state and local government respondents say that it can take months to identify sophisticated breaches. For the technology needed to fill in the gaps, the respondents most frequently listed end point/network security (58 percent), encryption (56 percent), threat intelligence (54 percent) and cyber-threat analytics (51 percent).

Public-service organizations need to integrate cyber defenses deeply into their organizations by employing a comprehensive end-to-end approach to digital security. As a first step, agencies should conduct a thorough assessment of their cybersecurity capabilities, while "pressure-testing" their defenses to determine whether they can withstand a targeted attack. They also need to identify and minimize their network exposure and focus on protecting priority assets. The following cybersecurity areas should be considered priorities for investment and greater leadership attention:

- **Governance:** Focus on accountability to nurture a cybersecurity-minded culture, measure and report cybersecurity performance, develop attractive cybersecurity incentives for employees, and create a clear-cut cybersecurity chain of command. Leaders need to redefine cybersecurity success as more than simply achieving compliance targets. Getting the right level of visibility and authority is critical to discovering and responding to threats in a timely manner.
- **Agency exposure:** Assess cybersecurity incident scenarios to understand those that could materially affect the organization. Identify key drivers, decision points and barriers to the development of remediation and transformation strategies.
- **Strategic threat context:** Drive the organization to explore specific cybersecurity threats, including an analysis of geopolitical risks, and to identify what cybersecurity-related activities and technologies similar organizations are undertaking and deploying. These steps will ensure that an agency's security program aligns with its overall strategy.
- **Cyber resilience:** Assess the organization's ability to deliver operational excellence in the face of disruptive cyber adversaries, and use "design for resilience" techniques to limit the impact of an attack.
- **Cyber response readiness:** Put in place a robust response plan, provide effective cyber incident escalation paths, and ensure solid stakeholder involvement across all agency functions. Test the ability of team members to cooperate during crisis-management incidents.

- **Investment efficiency:** Develop in-house expertise to drive smart cybersecurity investments and the most effective allocation of funding and resources. Compare organizational investments against benchmarks, organizational objectives and cybersecurity trends. Asset management can be difficult for government organizations, but this is a critical component of any security campaign.

Government agencies should approach cybersecurity with an organizational mindset—one capable of continually evolving and adapting to changing threats. State-of-the-art cybersecurity will require not only investments in innovation and training but also rock-solid commitment from leaders.

30. Security Pros Need a Mentor*

Here's Why and How

DANIEL J. LOHRMANN

Finding and keeping cybertalent is a top global concern for public-and private-sector organizations.

At the same time, security professionals understand the need to network, learn from case studies and gain a deeper level of professional interaction with cybersecurity experts and experienced leaders from a similar context or industry.

But in government, attracting, developing and enhancing the career effectiveness of security pros has become a crisis-level challenge.

So what is being done to help? How can new security leaders enhance their careers and learn from others who have gone before them? What alternatives are available to assist local and state government security professionals who may not have the same financial resources as their private-sector counterparts? How can experienced CISOs and other leaders give back to the community and leave a positive legacy—while expanding their own horizons at the same time?

Short answer: Mentoring.

More specifically: Find a mentor.

Or be a mentor.

Or both.

One excellent mentoring program has been running for almost six years

*Originally published as Daniel J. Lohrmann, "Security Pros Need a Mentor: Here's Why and How," *Government Technology*, February 11, 2018. Reprinted with permission of the publisher.

in state and local governments nationwide. Here's why it has been so successful and is a national best-practice, with personal stories to show the career and organizational benefits.

Background on the Multi-State Information Sharing and Analysis Center (MS-ISAC) Mentoring Program

I have written many articles on the development and benefits of active participating in your industry's information sharing and analysis center (ISAC). You can learn more about industry ISACs here and specifically the MS-ISAC here and here and here.

The MS-ISAC Mentoring program began in 2012, and this article describes the early launch of the mentoring program, which is still free to participants. *Note: Membership in the MS-ISAC is also free for government organizations.*

I spoke with Mike Aliperti, a friend and longtime leader at the MS-ISAC, about the current mentoring program. Mike is the Multi-State Information Sharing and Analysis Center (MS-ISAC) VP of Stakeholder Engagement for the Center for Internet Security (CIS). CIS is a nonprofit organization whose mission is to provide cyberthreat prevention, protection, response and recovery for the nation's state, local, territorial and tribal (SLTT) governments as well as private-sector communities.

Mike has been with CIS for over seven years and is responsible for oversight of all of the MS-ISAC Members. Mike provides leadership in developing programs, organizational and financial strategies to deliver services to MS-ISAC members. Mike is working to build on and enhance the great relationships between federal, state and local governments and private industry by sharing information and collaborating where possible.

Here's that interview.

Dan Lohrmann (DL): Can you briefly describe the MS-ISAC mentoring program in 2018? What does it entail?

Mike Aliperti (MA): The goal for the MS-ISAC Mentoring Program is to provide an opportunity for security leaders in management positions (chief information security officers and chief security officers) to network and learn from the experience of current security leaders. These professional partnerships, through regular communication, were to foster a trusted mentor/mentee relationships. The opportunity provides the mentee with a valued partner for problem-solving, career guidance, and insight into shared experiences and solutions.

The process is informal beside the quarterly calls with the full group, each pairing decides what goals are set for their relationship and how often they communicate. We encourage each to attend the MS-ISAC annual meeting so that they can meet F2F.

DL: Tell us your view on the benefits provided by the MS-ISAC Mentoring Program.

MA: We have learned that both the mentee and mentor gain from the relationship.

- Having someone as an adviser and someone to talk to about specific security issues and questions.
- Sharing successes and things not so successful and being able to learn from these.
- Help with educational opportunities for mentee.
- Knowledge of contributing to overall security programs of participants.
- Enjoyed getting to know an individual person with similar job responsibilities and sharing knowledge with them.
- Given the opportunity to share knowledge with someone who was new in the information security role.
- Provided a venue to establish peer-to-peer contacts.
- Learning more about other cyber security programs and how problems were addressed

DL: How many people are involved and how does it help your members improve in their roles?

MA: The number of participants has grown over the years, as you know we start with eight pairings and currently (for the 2018 cycle) there are 29 pairings with a possible additional six more. The biggest benefit, I believe, is that mentees see that there is (meaningful) help available. A number of mentees have become mentors. The more involved in the program they become, the more they understand what the MS-ISAC offers to them. A large number of mentors participate year after year in the program.

DL: How has the mentoring program evolved over the past five years since inception?

MA: We always have issues with some mentees looking for more technical support from their mentor, when the program is a security leadership mentoring program.

- We created a process to "rematch" if the pairing is not good.
- Created a Mentoring Program Guideline to assist the pairing in creating and maintaining a mentoring relationship as well as setting goals.

- Establish matching criteria. Match local person with local person. Match to similar types of experiences. Consider time zones when matching.
- Face-to-face introductions at annual meeting.

DL: Why do you think this program is growing?

MA: The MS-ISAC membership has grown from 200 members to over 1,950. The participants are spreading the word and encouraging others to join the program. Also, with the large turnover in chief information security officers and chief security officers, there is a very small number of state CISO that are the same from six years ago. Dan, the number of state CISOs from when you were one, I can count on one hand.

Final Thoughts

In a BrightTalk webcast, I moderated a panel with cyberindustry leaders on best practices regarding attracting and retaining cybertalent. One top trend that was recommended was finding a security-focused mentor or being a mentor for others in the industry.

In addition, the Information Systems Security Association offers this CISO Mentoring Webinar Series, which is also an excellent resource.

In my opinion, there is no career substitute for finding a mentor or being a mentor. While we all have seasons of our lives when this may not be possible, try to make it a priority. You'll be glad you did. Interested government security professionals should contact Tammie at the MS-ISAC for more details on how to get involved in mentoring.

I also believe that this mentoring program is a model for all industry sectors beyond government, and the other ISACs or professional organizations may want to consider this MS-ISAC best practice approach.

My hope is that this (admittedly long) blog will convince you to get involved and either find a mentor, or be a mentor or both.

31. Mecklenburg County Refuses to Pay Ransom to Cyber Hackers*

MARY ANN BARTON

Mecklenburg County, N.C., will not pay a $23,000 ransom after hackers demanded the amount after taking over some of the county's computer systems.

"This is cyber warfare," Commissioner Trevor Fuller said Dec. 8. "It highlights the world we live in. As the county gets more technology to deliver services more efficiently, it can present tremendous risks. But I'm glad we have backups. It gave us the option to not pay the ransom."

The county announced the breach, which took place Dec.4, on Dec. 5 via its Twitter account: "We are experiencing a computer-system outage. If you are planning to go to a County office to conduct business, please contact the office prior to going to ensure you can be served." The message included a link to more details about the ransomware attack.

"I am confident that our backup data is secure and we have the resources to fix this situation ourselves," Mecklenburg County Manager Dena R. Diorio said in a statement. "It will take time, but with patience and hard work, all of our systems will be back up and running as soon as possible."

Fuller said the county Board of Commissioners empowered Diorio to take the necessary steps to activate crisis plans for each department and to consult with third-party experts to get to the bottom of the ransomware attack.

---

*Originally published as Mary Ann Barton, "Mecklenburg County Refuses to Pay Ransom to Cyber Hackers," *NACo County News*, Dec. 8, 2017. Reprinted with permission of the publisher.

Mecklenburg County is not alone. In the first quarter of 2017, the most recent figures available, there have been 745 victims of ransomware, losing more than $512,000 to cyber hackers, the FBI said, along with much more lost in work hours. At that pace, the FBI could see more victims than last year, when 2,673 notified the crime-fighting agency about ransomware attacks.

The county's decision not to pay the ransom received national attention in a story published in *The New York Times*: "In a world rocked by hackers, trolls and online evildoers of all stripes, the good people of the internet have long looked for a hero who would refuse to back down. Finally, someone has said enough is enough. And that someone is the government of Mecklenburg County, N.C."

The county decided not to pay the ransom after discussing the issue with several third-party cyber security experts who told them that the time-frame for fixing the systems and dealing with the hackers would be about the same, Diorio said.

"It was going to take almost as long to fix the system after paying the ransom as it does to fix it ourselves," she said. "And there was no guarantee that paying the criminals was a sure fix."

After the county announced its decision not to pay, the hackers tried again to invade county systems, several times, Fuller said. "We've been repelling them so far. It is a bizarre situation to be in. You hear about it in the news but to experience it is like this..."

In an email to county workers, Diorio wrote: "As a result of our decision not to pay the ransom, ITS (Information Technology Services) is reporting that the cyber criminals are redoubling their efforts to penetrate the County's systems, primarily through emails that contain fraudulent attachments with viruses that could further damage our systems."

The county temporarily disabled the ability to open attachments from file services such as Dropbox or Google Docs. In her email, Diorio also addressed employees directly about the incident: "I also want to reiterate that the County is the victim in this situation and that no individual employee should feel responsible for this incident."

The initial attack took place after an employee opened an email and clicked on an attachment, which triggered a program called "LockCrypt" that spread encrypted data across 48 of the county's 500 servers, Diorio noted at a news conference. The county shut down other servers to protect them. A note from the hackers read: "Your information is locked," and gave the county instructions on how to pay the ransom of $23,000 or two Bitcoins.

The county got the word out to the public about the hack via the news conference (aired live on the county's Facebook page), social media and local radio interviews. Diorio's news conference Wednesday included all department heads, who were made available to answer any questions about services hampered by the attack.

County offices remain open to serve the public and the county said it would use backup data to rebuild compromised applications. The top priorities were health and human services, the court system, land use and environmental services. Other county offices impacted by the ransomware attack included the tax office, register of deeds, assessor's office, park and recreation, child support enforcement and finance.

The county has asked its employees and residents to be patient during the time it takes to get all the systems back up and running. The county asked residents to call ahead if they have business with the county. It's estimated all systems will be back up by Dec. 31.

The county is consulting with the governor's office, the FBI, the Secret Service, Department of Homeland Security, and local business leaders in the field. "We're gratified for those who stepped up to help," Fuller said, adding that the county will conduct a thorough investigation into how the attacks occurred once its systems are back up, to check for vulnerabilities.

For any county thinking of taking a similar stand, be sure to be prepared.

If the county didn't have backups, "we would have had to pay, no matter what," said Commissioner Jim Plunkett. "If you do not have a clean backup, you are completely at risk. They can shut the county down."

32. Fighting Fake News*

MARCUS BANKS

Librarians—whether public, school, academic, or special—all seek to ensure that patrons who ask for help get accurate information.

Given the care that librarians bring to this task, the recent explosion in unverified, unsourced, and sometimes completely untrue news has been discouraging, to say the least. According to the Pew Research Center, a majority of U.S. adults are getting their news in real time from their social media feeds. These are often uncurated spaces in which falsehoods thrive, as revealed during the 2016 election. To take just one example, Pope Francis did not endorse Donald Trump, but thousands of people shared the "news" that he had done so.

Completely fake news is at the extreme end of a continuum. Less blatant falsehoods involve only sharing the data that puts a proposal in its best light, a practice of which most politicians and interest group spokespeople are guilty.

The news-savvy consumer is able to distinguish fact from opinion and to discern the hallmarks of evasive language and half-truths. But growing evidence suggests that these skills are becoming rarer. A November 2016 study by the Stanford History Education Group (SHEG) showed that students have difficulty separating paid advertising from news reporting, and they are apt to overlook clear evidence of bias in the claims they encounter. These challenges persist from middle school to college.

According to SHEG Director Sam Wineburg, professor at Stanford Graduate School of Education, "nothing less than our capacity for online civic reasoning is at risk."

---

*Originally published as Marcus Banks, "Fighting Fake News," *American Libraries*, December 27, 2016. Reprinted with permission of the publisher.

Librarians and Journalists: Natural Allies

Librarians can help change this trend. "Librarians are natural allies for educators in helping students become critical news consumers," says Wineburg. The profession's deep commitment to verified sources and reliable information mirrors similar values—accountability for accuracy, careful research before drawing firm conclusions, and a willingness to correct errors—that drive responsible journalism.

One emerging solution among journalists is the Trust Project, an initiative of the Markkula Center for Applied Ethics at Santa Clara (Calif.) University.

Headed by longtime reporter Sally Lehrman, director of Santa Clara's journalism ethics program, the Trust Project has partnered with nearly 70 media organizations to develop a collection of color-coded digital "Trust Indicators" that signify reliable and responsible reporting. Indicators include a commitment to seeking diverse perspectives, linking out to credible sources of further information, offering clear markers regarding whether an article presents opinion or news, and providing information about an article's author. The complete set is available at the Trust Project website.

Still in the works for the project is computer code that will allow partner media organizations to note when they have achieved a Trust Indicator, which serves as a proxy for reliable journalism. This code should be broadly available by mid-2017. Services such as Facebook and Google would surface these materials more prominently in news feeds and search results, while readers would see clear visual icons that demonstrate fulfillment of the Trust Indicators. As Lehrman explains, "These icons would be cognitive shortcuts to route readers to more reliable sources of news."

She also notes a strong desire by consumers to be active participants in the shaping of the news, rather than merely a passive audience. In that spirit, she welcomes input and feedback from librarians about how to best achieve the aims of the Trust Project.

Direct collaboration with journalists is another route to increasing media literacy. For example, the Dallas Public Library (DPL) will host an eight-week training course in community journalism for high school students. Its "Storytellers without Borders" project, one of the winners of the 2016 Knight News Challenge, includes oversight from professional librarians as well as reporters at the *Dallas Morning News*.

Students will rotate among three DPL branch locations that represent the socioeconomic and cultural diversity of the city. Journalists will mentor students on how to ask focused questions, while librarians will describe how to use research databases to find accurate information. Library staffers will also provide instruction on how to use multimedia editing tools. In April

2017 these budding digital journalists, with their new skills in the art of providing credible and engaging content, will showcase their efforts at the Dallas Book Festival.

Information Literacy at Your Library

The Trust Project and "Storytellers without Borders" are high-profile efforts, but any library can lead educational programs about the importance of media literacy.

As the SHEG study reveals, this training should begin with young students and continue through college. Resources that range from free LibGuides to enhanced school curricula are available for libraries around the country.

Librarians at Indiana University East in Richmond have developed a LibGuide about how to identify fake news, complete with detailed images of what questions to ask while perusing a site. The News Literacy Project, founded by former *Los Angeles Times* reporter Alan Miller, offers a comprehensive curriculum of classroom, after-school, and e-learning programs for middle and high school students; the Center for News Literacy at Stony Brook (N.Y.) University offers similar resources for teaching college students.

Despite the clear need for increased media literacy, one risk is that this topic will always be perceived as optional—nice to know but not essential. Wineburg argues that this is misguided. "Online civic literacy is a core skill that should be insinuated into the warp and woof of education as much as possible," he says. In a paper for *College & Research Libraries News*, Brian T. Sullivan, information literacy librarian, and Karen L. Porter, sociology professor, of Alfred (N.Y.) University map out how to convert those one-shot information literacy training sessions into full programs with embedded librarians.

Librarians can play a vital role in helping everyone, of any age, become critical and reflective news consumers. One positive outcome of the current furor about fake news may be that information literacy, for media and other types of content, will finally be recognized as a central skill of the digital age.

33. How Airplane Crash Investigations Can Improve Cybersecurity*

SCOTT SHACKELFORD

While some countries struggle with safety, U.S. airplane travel has lately had a remarkable safety record. In fact, from 2014 through 2017, there were no fatal commercial airline crashes in the U.S.

But those years were fraught with other kinds of trouble: Security breaches and electronic espionage affected nearly every adult in the U.S., along with the power grid in Ukraine and the 2016 U.S. presidential campaign, to name a few. As a scholar of cybersecurity policy, I think it's time that my own industry took some lessons from one of the safest high-tech transportation methods of the 21st century.

Like today in cybersecurity, the early days of U.S. air travel weren't regulated particularly closely. And there were a huge number of accidents. Only after public tragedies struck did changes occur. In 1931, a plane crash in Kansas killed legendary Notre Dame football coach Knute Rockne. And in 1935, U.S. Sen. Bronson Cutting of New Mexico died in the Missouri crash of TWA flight 6. These events helped contribute to the 1938 creation of the first U.S. Air Safety Board. But it took until 1967 for the new Department of Transportation to be created with an independent National Transportation Safety Board.

Since then, the NTSB has rigorously investigated all airplane crashes and other transportation incidents in the U.S. Its public reports about its

*Originally published as Scott Shackelford, "How Airplane Crash Investigations Can Improve Cybersecurity," *The Conversation*, Feb 21, 2018. Reprinted with permission of the publisher.

findings have informed changes in government regulations, corporate policies and manufacturing standards, making air travel safer in the U.S. and around the world.

As cybersecurity incidents proliferate around the country and the globe, businesses, government agencies and the public shouldn't wait for an inevitable disaster before investigating, understanding and preventing these failures. Nearly a century after the original Air Commerce Act in 1926, calls, including my own, are mounting for the information industry to take a page from aviation and create a cybersecurity safety board.

The Flight Plan to Safer Skies

The creation of the National Transportation Safety Board was the first independent agency charged with investigating the safety of various transportation systems, from highways and pipelines to railroads and airplanes. Since 1967, the NTSB has investigated more than 130,000 accidents.

These investigations are vital since they help establish "the who, what, where, when, how and [perhaps] why behind an incident." After the facts are determined, policymakers can back up, and often have backed up, NTSB recommendations with new regulations. Failing that, it is common for air carriers, for example, to voluntarily implement changes it suggests. A similar approach could help improve the internet, a new technology that, like airplanes, is tying the world closer together even as it threatens our shared security.

The Case for a Cybersecurity Safety Board

Two elements of the NTSB may be particularly useful for enhancing cybersecurity. First, it separates fact-finding proceedings from any questions of legal liability. Second, these investigations are broad, involving various stakeholders like manufacturers and airline companies. Cyberspace is similarly made up of a wide range of companies and technologies.

A cybersecurity safety board need not in fact be national. It could begin from the bottom up, with companies partnering together to protect their customers by sharing best practices.

Critics of establishing a cybersecurity safety board would likely contend that the speed at which technologies change makes it difficult for any recommendations, even if they were quickly implemented, to sufficiently protect organizations from cyber attacks. NTSB investigations can take a year or more; to ensure findings were still relevant, cybersecurity inquiries would need to be faster, such as by streamlining cyber forensics and relying on widely used tools such as the National Institute for Standards and Technology Cybersecurity Framework.

Other challenges include standardizing terminology across the industry and identifying the right experts to look into data breaches, which might be easier said than done given the talent shortage among cybersecurity professionals. Broad-based cybersecurity educational programs, like a new partnership between the law, business and computer science schools here at Indiana University, should be encouraged to help address this shortfall.

A Path Forward

Additional measures would likely be required to make a cybersecurity safety board successful, such as launching investigations only for serious breaches like those involving critical infrastructure.

More nations and regions—including the European Union—are imposing stringent requirements on companies that suffer data breaches, including mandatory reporting of cyberattacks within 72 hours and more rigorous preventive measures. Businesses, governments and scholars around the world are working on how to improve data security. If they came together to support a global network of cybersecurity safety boards, their efforts could promote cyber peace for people and institutions alike.

All that is needed is the will to act, the desire to experiment with new models of cybersecurity governance and the recognition that we should learn from history. As President Franklin D. Roosevelt famously said, "It is common sense to take a method and try it: If it fails, admit it frankly and try another. But above all, try something."

34. National Cybersecurity Workforce Framework*

U.S. DEPARTMENT
OF HOMELAND SECURITY

Overview

Cybersecurity is a national priority and critical to the well-being of all organizations. As technology becomes increasingly more sophisticated, demand for an experienced and qualified workforce has never been higher. Large and small organizations from both the public and private sectors are creating new cyber jobs and hiring thousands of cyber professionals to protect networks and information systems.

Organizations must have the right cyber staff in place to protect their business. The Department of Homeland Security (DHS) is committed to strengthening the workforce to ensure that organizations have the information and tools needed to protect their business and meet the challenges of the future.

DHS has the programs and capabilities to help your organization build a world-class cyber workforce:

- Identify and quantify your cybersecurity workforce
- Understand workforce needs and skills gaps
- Hire the right people for clearly defined roles
- Enhance employee skills with training and professional development
- Create programs and experiences to retain top talent

*Public document originally published as U.S. Department of Homeland Security, "National Cybersecurity Workforce Framework," https://www.dhs.gov/national-cybersecurity-workforce-framework.

The National Initiative for Cybersecurity Education (NICE) Cybersecurity Workforce Framework provides a blueprint to categorize, organize, and describe cybersecurity work into Categories, Specialty Areas, Work Roles, tasks, and knowledge, skills, and abilities (KSAs). The NICE Framework provides a common language to speak about cyber roles and jobs and helps define personal requirements in cybersecurity.

Categories, Specialty Areas and Work Roles

Within the NICE Framework, there are seven Categories, each comprising of several Specialty Areas. Additionally, within each Specialty Area, there are a set of Work Roles. Each Work Role has Knowledges, Skills and Abilities (KSAs) required for the role, as well as Tasks performed by the role. This organizing structure is based on extensive job analyses that groups together work and workers that share common major functions, regardless of job titles or other occupational terms.

1. Analyze

Performs highly-specialized review and evaluation of incoming cybersecurity information to determine its usefulness for intelligence.

All-Source Analysis. Analyzes threat information from multiple sources, disciplines, and agencies across the Intelligence Community. Synthesizes and places intelligence information in context; draws insights about the possible implications.

Exploitation Analysis. Analyzes collected information to identify vulnerabilities and potential for exploitation.

Language Analysis. Applies language, cultural, and technical expertise to support information collection, analysis, and other cybersecurity activities.

Targets. Applies current knowledge of one or more regions, countries, non-state entities, and/or technologies.

Threat Analysis. Identifies and assesses the capabilities and activities of cybersecurity criminals or foreign intelligence entities; produces findings to help initialize or support law enforcement and counterintelligence investigations or activities.

2. Collect and Operate

Provides specialized denial and deception operations and collection of cybersecurity information that may be used to develop intelligence.

Collection Operations. Executes collection using appropriate strategies and within the priorities established through the collection management process.

Cyber Operational Planning. Performs in-depth joint targeting and cybersecurity planning process. Gathers information and develops detailed Operational Plans and Orders supporting requirements. Conducts strategic and operational-level planning across the full range of operations for integrated information and cyberspace operations.

Cyber Operations. Performs activities to gather evidence on criminal or foreign intelligence entities to mitigate possible or real-time threats, protect against espionage or insider threats, foreign sabotage, international terrorist activities, or to support other intelligence activities.

3. Investigate

Investigates cybersecurity events or crimes related to information technology (IT) systems, networks, and digital evidence.

Cyber Investigation. Applies tactics, techniques, and procedures for a full range of investigative tools and processes to include, but not limited to, interview and interrogation techniques, surveillance, counter surveillance, and surveillance detection, and appropriately balances the benefits of prosecution versus intelligence gathering.

Digital Forensics. Collects, processes, preserves, analyzes, and presents computer-related evidence in support of network vulnerability mitigation and/or criminal, fraud, counterintelligence, or law enforcement investigations.

4. Operate and Maintain

Provides the support, administration, and maintenance necessary to ensure effective and efficient information technology (IT) system performance and security.

Customer Service and Technical Support. Addresses problems; installs, configures, troubleshoots, and provides maintenance and training in response to customer requirements or inquiries (e.g., tiered-level customer support). Typically provides initial information to the Incident Response (IR) Specialty.

Data Administration. Develops and administers databases and/or data management systems that allow for the storage, query, protection, and utilization of data.

Knowledge Management. Manages and administers processes and tools that enable the organization to identify, document, and access intellectual capital and information content.

Network Services. Installs, configures, tests, operates, maintains, and manages networks and their firewalls, including hardware (e.g., hubs, bridges, switches, multiplexers, routers, cables, proxy servers, and protective distributor systems) and software that permit the sharing and transmission of all spectrum transmissions of information to support the security of information and information systems.

Systems Administration. Installs, configures, troubleshoots, and maintains server configurations (hardware and software) to ensure their confidentiality, integrity, and availability. Manages accounts, firewalls, and patches. Responsible for access control, passwords, and account creation and administration.

Systems Analysis. Studies an organization's current computer systems and procedures, and designs information systems solutions to help the organization operate more securely, efficiently, and effectively. Brings business and information technology (IT) together by understanding the needs and limitations of both.

5. Oversee and Govern

Provides leadership, management, direction, or development and advocacy so the organization may effectively conduct cybersecurity work.

Cybersecurity Management. Oversees the cybersecurity program of an information system or network, including managing information security implications within the organization, specific program, or other area of responsibility, to include strategic, personnel, infrastructure, requirements, policy enforcement, emergency planning, security awareness, and other resources.

Executive Cyber Leadership. Supervises, manages, and/or leads work and workers performing cyber and cyber-related and/or cyber operations work.

Legal Advice and Advocacy. Provides legally sound advice and recommendations to leadership and staff on a variety of relevant topics within the pertinent subject domain. Advocates legal and policy changes, and makes a case on behalf of client via a wide range of written and oral work products, including legal briefs and proceedings.

Program/Project Management and Acquisition. Applies knowledge of data, information, processes, organizational interactions, skills, and analytical expertise, as well as systems, networks, and information exchange capabilities to manage acquisition programs. Executes duties governing hardware, software, and information system acquisition programs and other program management policies. Provides direct support for acquisitions that use information technology (IT) (including National Security Systems), applying IT-related laws and policies, and provides IT-related guidance throughout the total acquisition life cycle.

Strategic Planning and Policy. Develops policies and plans and/or advocates for changes in policy that support organizational cyberspace initiatives or required changes/enhancements.

Training, Education and Awareness. Conducts training of personnel within pertinent subject domain. Develops, plans, coordinates, delivers and/or evaluates training courses, methods, and techniques as appropriate.

6. Protect and Defend

Identifies, analyzes, and mitigates threats to internal information technology (IT) systems and/or networks.

Cyber Defense Analysis. Uses defensive measures and information collected from a variety of sources to identify, analyze, and report events that occur or might occur within the network to protect information, information systems, and networks from threats.

Cyber Defense Infrastructure Support. Tests, implements, deploys, maintains, reviews, and administers the infrastructure hardware and software that are required to effectively manage the computer network defense service provider network and resources. Monitors network to actively remediate unauthorized activities.

Incident Response. Responds to crises or urgent situations within the pertinent domain to mitigate immediate and potential threats. Uses mitigation, preparedness, and response and recovery approaches, as needed, to maximize survival of life, preservation of property, and information security. Investigates and analyzes all relevant response activities.

Vulnerability Assessment and Management. Conducts assessments of threats and vulnerabilities; determines deviations from acceptable configurations, enterprise or local policy; assesses the level of risk; and develops and/or recommends appropriate mitigation countermeasures in operational and nonoperational situations.

7. Securely Provision

Conceptualizes, designs, procures, and/or builds secure information technology (IT) systems, with responsibility for aspects of system and/or network development.

Risk Management. Oversees, evaluates, and supports the documentation, validation, assessment, and authorization processes necessary to assure that existing and new information technology (IT) systems meet the organization's cybersecurity and risk requirements. Ensures appropriate treatment of risk, compliance, and assurance from internal and external perspectives.

Software Development. Develops and writes/codes new (or modifies existing) computer applications, software, or specialized utility programs following software assurance best practices.

Systems Architecture. Develops system concepts and works on the capabilities phases of the systems development life cycle; translates technology and environmental conditions (e.g., law and regulation) into system and security designs and processes.

Systems Development. Works on the development phases of the systems development life cycle.

Systems Requirements Planning. Consults with customers to gather and evaluate functional requirements and translates these requirements into technical solutions. Provides guidance to customers about applicability of information systems to meet business needs.

Technology R&D. Conducts technology assessment and integration processes; provides and supports a prototype capability and/or evaluates its utility.

Test and Evaluation. Develops and conducts tests of systems to evaluate compliance with specifications and requirements by applying principles and methods for cost-effective planning, evaluating, verifying, and validating of technical, functional, and performance characteristics (including interoperability) of systems or elements of systems incorporating IT.

35. Eliminating Network Blind Spots and Preventing Breaches*

Reggie Best

Each year a growing number of critical cyber incidents are discovered in government systems and networks. Most often, these incidents are reported only after significant damage has been done and critical, secret or personally identifiable data has been compromised or exfiltrated. In addition, there has been a significant rise in ransomware attacks, as evidenced by this year's highly public examples, WannaCry and NotPetya.

And as the number of attacks increases, so does their sophistication, making it difficult to ensure networks are properly secured while still providing availability to critical data and systems. It's a challenging balance for government agencies, but protecting networks, systems, and information while continually providing essential services to the public is achievable.

But while this balance can be struck, it's important to consider the myriad security threats facing government networks and to remember that they contain highly confidential, sensitive or proprietary information. As government organizations increasingly move to the cloud, their networks become more complicated and vulnerable with third-party connections and internet-of-things devices greatly increasing the attack surface. And now that the federal government is using drones for missions such as disaster relief, law enforcement, border security, military training and more, the threat vector and surface become much wider.

Agencies must find a way to monitor the entire environment, from end-

*Originally published as Reggie Best, "Eliminating Network Blind Spots and Preventing Breaches," *GCN Magazine*, Aug 10, 2017. Reprinted with permission of the publisher.

points and across physical network infrastructure to the cloud. This means spotting questionable or suspicious dynamic infrastructure changes, potential leak paths to the internet, unknown devices and shadow IT infrastructure.

After all, how can you secure something if you can't see it?

That's where cyber situational awareness is critical. Agencies must have a real-time, holistic view of known and unknown threats to the infrastructure as they emerge and change so they can identify threats and vulnerabilities and develop effective responses to an attack.

But today's increasingly connected world is disrupting the traditional thinking about networks and how to properly secure systems. When we move to non-proprietary communication and network technologies and use more off-the-shelf commercial operating systems, we open them to additional cyber risk. These newly exposed risks can give malicious actors an entrance into the entire infrastructure.

Take the recent NotPetya cyberattack in the Ukraine that essentially paralyzed of the country's computer systems. Attackers used a software vulnerability as their gateway to infect Ukrainian government computers with ransomware. The lack of endpoint and network visibility and context and the inability to understand attacker activity on the network shows how simple OS vulnerabilities can be have catastrophic consequences that can spread incredibly quickly through government networks.

Implementing Cyber Situational Awareness

A real-time network situational awareness capability is critical to ensure the stability of government and critical infrastructure operations. It combines a deep assessment of the current network security operations to identify potential weak areas or vulnerabilities with a plan to detect and mitigate threats.

With cyber situational awareness, agencies can:

- Discover network segments and endpoints, unknown rogue devices and shadow IT infrastructure.
- Identify potential leaky paths that attackers can use to explore the network for vulnerabilities and access sensitive data.
- Detect unauthorized communication attempts to external servers for the purposes of installing additional malicious software or attempting command and control of internal systems.
- Find misconfigurations or network segmentation problems that could create risk or become vulnerable to attack.

- Discover newly inserted, possibly rogue wireline or wireless infrastructure devices, firewalls, routers or other network functions (e.g., virtualized) acting as packet forwarders.
- Detect any data exfiltration from the network to malware servers.

Government cybersecurity thought leaders increasingly acknowledge that depending on perimeter defenses and endpoint-centric protections won't cut it—it hasn't so far. They assume that malicious actors will breach the perimeter. Instead of putting their heads in the sand hoping endpoint defenses will adequately protect the organization, sophisticated security teams are looking to detect malicious, anomalous behavior on the network infrastructure itself. They understand that real-time detection and proactive remediation will provide better results.

Citizens rely on and trust government agencies to protect the nation's infrastructure and mission-critical information. Once a threat accesses the network, it's too late, the damage is done. Agencies that are using cyber situational awareness have real-time and accurate visibility needed to properly protect their networks and to keep our infrastructure safe.

36. The Cloud and Enterprise Cybersecurity*

Leveling the Playing Field

MICROSOFT

Intelligent security is something we all strive for in the ever-evolving world of cyber threats to the enterprise. The status quo is no longer able to keep up with the pace at which threats morph and replicate themselves across the globe. The "bad guys," in many cases, are better funded and staffed with greater cyber expertise than the agencies which they target. Another growing concern is that the networks from which these global assaults are launched are almost always larger and more resilient than the network they are attempting to compromise.

What is a CISO to do? What is the hope of the enterprise when the traditional methodologies of cyber security fail or are prone to compromise themselves? The answer? Enterprises must go global with their effort to not just defend but fight back.

Is an organization able to find the signal in the noise of data points? This we know: attackers aren't going to wait for security software to catch up. Industry reports show advanced cyber attacks can go undetected for approximately 200 days. In today's threat environment, organizations need intelligent security solutions that continually evolve to keep up with the latest threats as they emerge.

Using machine learning to detect advanced cyber attacks, a progressive, data-driven model of cybersecurity has emerged to speed up detection time

---

*Originally published as Microsoft, "The Cloud and Enterprise Cybersecurity: Leveling the Playing Field," Chapter 4, in *Cybersecurity: Protecting Local Government Digital Resources* (Washington, DC: ICMA and Microsoft, 2017). Reprinted with permission of the publisher.

and reduce risk. Where is this model playing out, one might ask? The model is alive, well, and delivering results in the Microsoft cloud.

Modus Operandi: The Advanced Attack at Work

When security professionals detect a breach, it's almost certain that the attacker has been active in the victim's environment for some time. But how long?

For many in the industry, "200 days" has been accepted as a standard to frame the average. But this "standard" is also problematic for a couple of reasons.

First, that's a long time. It's roughly six-and-a-half months that a sophisticated cyber attacker or syndicate has been at work inside the system. What does an advanced attack do for those 200 days after it's gained entry to the network? Today, attackers employ a mix of methods, using traditional techniques alongside new ones as they constantly explore ways to exploit both people and technologies. The longer an undetected attack lives in your system, the more intel it can glean, underscoring the importance of early detection. Throughout this dark and exposed time, an organization's sensitive data and intellectual property have been potentially exposed, moving closer to inevitable compromise.

The fear of what goes on during those 200 days has made this statistic a yardstick for CISOs, CSOs, and even CEOs. Today, companies, security professionals, and the tech industry at large are thirsty for new, more advanced security measures to drive that number down.

Second, CISOs and CSOs know that the number of days isn't the most important element of a breach. So, as a practical matter, "200 days" is just a milestone, a figure used to measure and discuss the industry's progress. Even one day is too long, and by the time it is discovered, it's always too late. Shrinking that number to zero is the ultimate goal.

To do that, organizations need a more intelligent approach to detect threats earlier and turn the tide against sophisticated cyberattacks. This essay is designed to give readers a glimpse into how advanced threats are working to compromise sensitive information, and how the advanced computing power of the cloud, combined with data science and human experts, can help reduce the time it takes for an organization to detect an attack.

Attackers that deploy advanced exploits are a constant concern for the small agency or the largest enterprise, and repercussions of an attack go well beyond the initial costs of a breach. Highly skilled, well-funded, and constantly evolving, these perpetrators have motives that range from theft, to industrial espionage, to full-blown nation-state attacks.

Risk: Exposure Is Steeper Than Ever. First, there are the financial concerns. Today, the many malicious actors and authors that utilize advanced attacks are looking to profit from their efforts. It's no surprise, then, that the damages keep going up.

In 2015, a new threshold was reached when a sophisticated attack ring successfully breached more than one hundred banks across thirty countries, with losses estimated to exceed $1 billion. Because of the heightened risk, cyber insurance policies are becoming a new operating expense for many companies, with premiums for that emerging offering set to triple by 2020, approaching $7.5 billion.

There are also the less quantifiable and potentially costlier scars that successful cyberattacks leave, such as damaged brands, wary customers, stagnant growth, and compromised diplomatic relations. While not directly attributable to a dollar sign, these impacts can have lasting negative effects on an organization: driving down customer loyalty, driving up public skepticism, and ultimately impacting security operations staff who must be held accountable for breaches.

Other attacks are motivated not by financial incentives, but by a quest for sensitive information. Take STRONTIUM, for example. STRONTIUM is a well-known activity group whose targets include government bodies, diplomatic institutions, journalists, and military forces.

It is not after money and doesn't care about size of a target. It is after the most sensitive data it can find. Similarly, the Red October attack group uncovered in 2013 was found to have been infiltrating government and diplomatic institutions for at least five years.

Although it sounds like something out of a spy novel, it's a real issue. Unseen costs of security breaches are something that even two decades ago would sound like the plot of a sci-fi story. With so much at stake, it's no wonder that budgets are increasing, and companies are hungry for new solutions to address the growing problem of advanced cyber attacks.

Nearly 80 percent of cyber attacks begin with a good old-fashioned con job, using spear phishing attacks with compelling ruses to get users to compromise their information. But as security provider McAfee noted, more sophisticated attacks are on the rise, including new integrity attacks that can modify internal processes and reroute data as it flows through the network.[1]

This was the technique used in that $1-billion bank heist. Attackers continue to evolve with new forms of malware that can better hide from detection or erase themselves altogether. Attack vectors are also changing: No longer content with targeting PCs and servers residing in the corporate headquarters, attackers look to compromise satellite offices; workers' home computers; and even the software inside of cell phones, wearable devices, and automobiles.

The Cyber Kill Chain: A Basic Understanding

Breaches generally involve six clear phases, known in the security intelligence community as the Cyber Kill Chain®, a phrase trademarked by Lockheed Martin. These phases can occur sequentially, in parallel, or in a different order altogether, and each also offers an opportunity to gain intelligence to defeat attackers.

A Proactive Security Model: Staying One Step Ahead

Due to the stealth nature of advanced attacks, companies must shift to a more proactive security model that focuses on improving their ability to sniff out the attacker and stop him in his tracks.

Whereas the traditional model of enterprise security began with protecting the network perimeter, experts now suggest a more proactive approach that begins with detection enabled by robust security analytics.

This model promotes a constantly improving cycle, as pre-breach defenses are continually improved with new intelligence from post-breach detection and response.

For the past few years, CISOs and CSOs have been working to make this shift by implementing security intelligence measures that use data and analytics in an effort to rapidly detect the next attack and improve defenses overall. This includes steps such as the following:

- Investing in advanced security software and secure hardware
- Training employees on security imperatives and risks
- Deploying a security intelligence event management (SIEM) solution
- Subscribing to (often multiple) threat intelligence feeds
- Developing processes to correlate threat data, and even hiring data scientists to analyze it.

Thus far, these tools and processes have comprised the bulk of the industry's response to advanced attacks. Like many early-stage efforts in the tech industry, they have had mixed results.

It's not that they aren't effective. Mandiant's 2016 M-Trends report shows that when companies are successful at detection using their own systems, the time of an advanced attack's residency is cut drastically.

But there are also complaints—including the expense, cumbersome integration, and the inefficient manual process of correlating threat data and feeding it into the system. And once everything is in place with the SIEM, there's another problem—noise. There are simply too many alerts, too much data, for even the most advanced enterprises to make sense of it all. If the goal of all these efforts is to shorten those 200 days to near real time, then cutting through the noise has become a major roadblock, and part of what keeps detection a (costly) step behind.

To keep up with advanced attacks, organizations should continue investing in their SIEMs and associated process. Only the cloud can offer next-generation protection, detection and remediation at the scale needed today—including alert mechanisms integrated through platform sensors—in a way that constantly evolves to improve protections with true security intelligence.

Improving Detection: The Importance of Clear Signal

Reducing the time it takes to detect an attack presents enterprises with a new dilemma: having too much security-related data to process yet still not having enough information to separate the signal from the noise and understand an incident quickly.

The challenge here is not just sheer volume, but also separation. Many indicators of attack either seem innocent on their own, or are separated by industries, distances, and time frames. Without clear insight into the whole data set, early detection becomes a game of chance. Even the largest enterprises are facing these limitations:

- Real threat intelligence requires more data than most organizations can acquire on their own.
- Finding patterns and becoming smarter in that huge data pool require advanced techniques like machine learning along with massive computing power.
- Ultimately, applying new intelligence so that security measures and technologies constantly improve requires human experts who can understand what the data are saying, and take action.

This is where Microsoft is working to turn the tide. As a platform and services company, Microsoft's threat and activity data come from all points in the technology chain, across every vertical industry, all over the world. Microsoft's security products and cloud technologies are designed to work together to report malicious threat data as problems occur. This provides a "flight data recorder" that enables us to diagnose attacks, reverse-engineer advanced threat techniques, and apply that intelligence across the platform.

Advanced machine learning and intelligence gathering techniques working with traditional security methodologies to provide a holistic, dynamic, flexible, and easily manageable enterprise cybersecurity posture. This approach leverages cloud-based technologies, which a customer simply would be unable to duplicate on its premises at the scale required to be of value.

From Months to Minutes: Applied Analytics and Continuous Improvement

For nearly two decades, Microsoft has been turning threats into useful intelligence that can help fortify its platform and protect customers. Since the Security Development Lifecycle born from early worm attacks like Blaster, Code Red, and Slammer, to modern security services woven into our platforms and services, the company has continually built processes, technologies, and expertise to detect, protect, and respond to evolving threats. As our digital environment increasingly dominates day-to-day lives, the importance of providing strong data protection and cybersecurity is apparent. From small towns to large cities, all local governments are susceptible to cyber attacks and need to take proactive steps to prevent and mitigate damage from attacks.

NOTE

1. McAfee Security, "McAfee Labs 2017 Threats Predictions November 2016" https://www.mcafee.com/us/resources/ reports/rp-threats-predictions-2017.pdf

37. Government Data in the Cloud*

Provider and User Responsibilities

SUBRATA CHAKRABARTI

Data breaches at the IRS, Office of Personnel Management, Securities and Exchange Commission and the State Department underscore the risk government agencies face with hacking on the rise. A recent study indicates that tight budgets may be part of the problem. Federal agencies have struggled to secure funds to keep data safe, though help may be on the way. As modernization funding becomes available, government IT leaders will need to make decisions about security.

Cloud service providers (CSPs) are also taking a fresh look at how to protect data, as the current rash of data breaches is unlikely to slow down. In fact, many cybersecurity experts expect 2018 to set new records for compromised information at organizations of all types. To protect data, CSPs and users alike must embrace their roles in preventing unauthorized access to sensitive information.

What CSPs Owe to the Organizations They Serve

Government IT managers should be able to trust the software-as-a-service vendors they use to handle sensitive data, and cloud providers must

*Originally published as Subrata Chakrabarti, "Government Data in the Cloud: Provider and User Responsibilities," *GCN Magazine*, May 16, 2018. Reprinted with permission of the publisher.

earn government user trust by deploying the most effective security measures available. It's not a one-off responsibility because hackers are actively devising new exploits, including weaponizing artificial intelligence, creating ransomware and testing internet-of-things endpoints for vulnerabilities.

Conscientious CSPs today offer sophisticated intrusion and exploit detection processes, and they conduct routine third-party scans and implement other standard security features. But the most forward-thinking providers are going above and beyond standard measures and using techniques such as "defense in depth" to safeguard data.

Also known as the "castle approach," defense in depth deploys multiple security mechanisms so that if one defense fails, another automatically takes its place. CSPs can apply this principle by deploying numerous protective layers for a systems that handle sensitive information, securing hardware, software and processes rather than concentrating on a single aspect. A comprehensive approach like defense in depth makes sense with threats coming from all directions.

What Government Organizations Need to Know About CSPs

Government IT professionals should be able to trust CSPs, but in an environment where hacking happens all too frequently, they have a responsibility to proactively ensure that their providers are using the latest techniques and technologies to protect data. Cloud service users should look for specific SaaS security features that address today's critical threats.

Multiple authentication options are a must-have feature on a cloud platform, including tools like Security Assertion Markup Language, an authorization protocol that lets agency administrators control authentication without requiring the CSP to store user passwords. Another must-have tool is access control that enables the organization's administrators to maintain separation of duties.

A bring-your-own-key solution (BYOK), meanwhile, may be a good choice for agencies with the most stringent security and compliance needs. With BYOK, organizations can manage their own encryption keys. This means they can encrypt and decrypt workspaces, maintaining sole access to their data in the cloud. Detailed audit logs of all encryption activity while using BYOK provide an organization with a comprehensive validation of system integrity.

To ensure CSPs are taking the necessary precautions with sensitive data, government IT professionals should confirm that the provider conducts third-party penetration tests periodically and is able to provide evidence of compliance with widely accepted standards, like certifications from the International Organization for Standardization or Service Organization Controls. CSPs with these certifications have passed a rigorous independent audit.

Data Safety Is a Two-Way Street

As endpoints proliferate and threats multiply, both CSPs and the organizations that use cloud services must proactively embrace their roles in keeping data safe. Cloud providers can do their part by continuously assessing their security posture and making sure the measures they take offer stringent protection to their customers, whether commercial or government.

Government IT professionals who oversee cloud services can do their part to protect data, assets and the communities they serve by proactively making sure their cloud solutions meet security standards. There's no end in sight to the hacking threat, but by understanding current security trends and innovations, IT professionals can make the right decisions.

38. Using Blockchain to Secure the "Internet of Things"*

Nir Kshetri

The world is full of connected devices—and more are coming. In 2017, there were an estimated 8.4 billion internet-enabled thermostats, cameras, streetlights and other electronics. By 2020 that number could exceed 20 billion, and by 2030 there could be 500 billion or more. Because they'll all be online all the time, each of those devices—whether a voice-recognition personal assistant or a pay-by-phone parking meter or a temperature sensor deep in an industrial robot—will be vulnerable to a cyberattack and could even be part of one.

Today, many "smart" internet-connected devices are made by large companies with well-known brand names, like Google, Apple, Microsoft and Samsung, which have both the technological systems and the marketing incentive to fix any security problems quickly. But that's not the case in the increasingly crowded world of smaller internet-enabled devices, like light bulbs, doorbells and even packages shipped by UPS. Those devices—and their digital "brains"—are typically made by unknown companies, many in developing countries, without the funds or ability—or the brand-recognition need—to incorporate strong security features.

Insecure "internet of things" devices have already contributed to major cyber disasters, such as the October 2016 cyberattack on internet routing company Dyn that took down more than 80 popular websites and stalled internet traffic across the U.S. The solution to this problem, in my view as a

Originally published as Nir Kshetri, "Using Blockchain to Secure the 'Internet of Things'" *The Conversation*, March 7, 2018. Reprinted with permission of the publisher.

scholar of "internet of things" technology, blockchain systems and cybersecurity, could be a new way of tracking and distributing security software updates using blockchains.

Making Security a Priority

Today's big technology companies work hard to keep users safe, but they have set themselves a daunting task: Thousands of complex software packages running on systems all over the world will invariably have errors that make them vulnerable to hackers. They also have teams of researchers and security analysts who try to identify and fix flaws before they cause problems.

When those teams find out about vulnerabilities (whether from their own or others' work, or from users' reports of malicious activity), they are well positioned to program updates and to send them out to users. These companies' computers, phones and even many software programs connect periodically to their manufacturers' sites to check for updates and can download and even install them automatically.

Beyond the staffing needed to track problems and create fixes, that effort requires enormous investment. It requires software to respond to the automated inquiries, storage space for new versions of software, and network bandwidth to send it all out to millions of users quickly. That's how people's iPhones, PlayStations and copies of Microsoft Word all stay fairly seamlessly up to date with security fixes.

None of that is happening with the manufacturers of the next generation of internet devices. Take, for example, Hangzhou Xiongmai Technology, based near Shanghai, China. Xiongmai makes internet-connected cameras and accessories under its brand and sells parts to other vendors.

Many of its products—and those of many other similar companies—contained administrative passwords that were set in the factory and were difficult or impossible to change. That left the door open for hackers to connect to Xiongmai-made devices, enter the preset password, take control of webcams or other devices, and generate enormous amounts of malicious internet traffic.

When the problem—and its global scope—became clear, there was little Xiongmai and other manufacturers could do to update their devices. The ability to prevent future cyberattacks like that depends on creating a way these companies can quickly, easily and cheaply issue software updates to customers when flaws are discovered.

A Potential Answer

Put simply, a blockchain is a transaction-recording computer database that's stored in many different places at once. In a sense, it's like a public bulletin board where people can post notices of transactions. Each post must be accompanied by a digital signature, and it can never be changed or deleted.

I'm not the only person suggesting using blockchain systems to improve internet-connected devices' security. In January 2017, a group including U.S. networking giant Cisco, German engineering firm Bosch, Bank of New York Mellon, Taiwanese electronics maker Foxconn, Dutch cybersecurity company Gemalto and a number of blockchain startup companies formed to develop just such a system.

It would be available for device makers to use in place of creating their own software update infrastructure the way the tech giants have. These smaller companies would have to program their products to check in with a blockchain system periodically to see if there was new software. Then they would securely upload their updates as they developed them. Each device would have a strong cryptographic identity, to ensure the manufacturer is communicating with the right device. As a result, device makers and their customers would know the equipment would efficiently keep its security up to date.

These sorts of systems would have to be easy to program into small devices with limited memory space and processing power. They would need standard ways to communicate and authenticate updates, to tell official messages from hackers' efforts. Existing blockchains, including Bitcoin SPV and Ethereum Light Client Protocol, look promising. And blockchain innovators will continue to find better ways, making it even easier for billions of "internet of things" devices to check in and update their security automatically.

The Importance of External Pressure

It will not be enough to develop blockchain-based systems that are capable of protecting "internet of things" devices. If the devices' manufacturers don't actually use those systems, everyone's cybersecurity will still be at risk. Companies that make cheap devices with small profit margins won't add these layers of protection without help and support from the outside. They'll need technological assistance and pressure from government regulations and consumer expectations to make the shift from their current practices.

If it's clear their products won't sell unless they're more secure, the unknown "internet of things" manufacturers will step up and make users and the internet as a whole safer.

39. How to Protect Patrons' Digital Privacy*

ANNE FORD

On April 3 President Trump signed a measure repealing Obama-era broadband privacy rules. Those rules, which had not yet gone into effect, would have required internet service providers (ISPs) to obtain customers' permission before selling their information to third parties—information that includes browsing history, location data, and other highly sensitive content.

"The librarian profession cares a great deal about the public's right to privacy, and this is a very serious erosion of that," says Alison Macrina. Macrina is director of Library Freedom Project, an organization that educates librarians about privacy-related rights and tools. "It means that whatever ISP we use in the library is now privy to the browsing habits of our patrons. It also means that a third party will be able to monetize the data of our community members."

The good news, she adds, is that in her view, "it is possible for us to have solutions to these problems, even if they're just interim technical solutions. I'm afraid that people, when they understand the immensity of these problems, will decide not to do anything, but doing something really does matter. It is possible to get the toothpaste back in the tube."

But how? Macrina and other privacy advocates offer several suggestions for libraries to consider as they determine how best to protect their patrons' digital privacy.

---

*Originally published as Anne Ford, "How to Protect Patrons' Digital Privacy," *American Libraries*, April 21, 2017. Reprinted with permission of the publisher.

Make Sure Your Library's Website Uses the HTTPS Protocol

Does your library website's URL begin with HTTPS, rather than just HTTP or WWW? Good. That means that communication between that website and a patron's computer is encrypted.

"If a patron [at home] uses a library's online catalog, for example, those communications include the queries they type into their browser, what they select, and in some cases, what they click on to download and read," says Marshall Breeding, founder of Library Technology Guides, editor of ALA TechSource's *Smart Libraries* newsletter, and author of several books including *Cloud Computing for Libraries* (ALA Tech Source, 2012).

"Without that HTTPS encryption, anyone can see what's going back and forth. It means that ISPs, as they peer from one part of the network to the other, can capture that traffic and sell it." With HTTPS, ISPs can see that a patron visited the library site, but not the pages they accessed or the books they looked up.

Right now, Breeding has found, fewer than 50 percent of the libraries he's queried use HTTPS for their websites, and fewer than 30 percent use it for their online catalog traffic. "These numbers are not great," he points out. "My concern is that library practices fall far short of the values that we have for protecting patron privacy."

For help implementing the HTTPS protocol, libraries can consult the Electronic Frontier Foundation or Let's Encrypt.

Negotiate with Your ISP

"Libraries are in a position, unlike individual consumers, to negotiate with their ISP to agree not to track users when they're using the library's internet connection, or at least not to sell user data," says Mike Robinson, professor of library science and head of the systems department at the Consortium Library of the University of Alaska Anchorage and Alaska Pacific University. "If it's a national ISP that's not making any exceptions, you might get caught up in that. But a lot of ISPs serve organizations, and I don't think libraries are going to be the only ones who don't want that tracking."

Consider Installing the Tor Browser on Library Computers

"Tor is a free and open source web browser that hides a lot of the information that is leaked when you use a typical web browser," Macrina explains.

"It hides your location info, and it obscures your browsing history from your ISP. If you use Tor, the ISP has nothing to sell."

That said, using Tor tends to mean that web pages will be downloaded more slowly. Users may also find themselves being required to enter CAPTCHAs to access some websites.

Educate Your Patrons About Virtual Private Networks (VPNs)

A VPN is a way to hide your online activity from your ISP. "It basically creates an encrypted tunnel for your Wi-Fi traffic to go through," Macrina says. The ISP can see that you're connected to the VPN, but it can't see what you're actually doing online.

Unfortunately, VPNs are difficult to use at scale in a library environment. "You have to have a lot of control over your users in a way that libraries don't," she says. Instead, "when we're teaching computer classes and things like that, we can recommend that patrons use VPNs on their own at home." Libraries can also educate patrons about the value of other privacy-protecting tools, such as the Disconnect and HTTPS Everywhere browser extensions.

Advocate with Lawmakers for Greater Online Privacy Protections

"All of these are technical solutions, but we need political solutions," Macrina says. "Libraries need to lobby on behalf of privacy legislation," such as the broadband privacy bills that are under way in Illinois and Minnesota. "We need to petition the lawmakers who voted badly on this and shame them for how much money they get from the telecommunications industry. We need to reach out to lawmakers who are sympathetic to privacy concerns and demand that they do something to reverse this. It's possible to have an internet where we don't have to sacrifice all this information just to get basic services."

40. Monitoring Employees' Use of Company Computers and the Internet*

Texas Workforce Commission

Business-related use of the Internet has grown by leaps and bounds in the last few years. At the same time, more and more employees must use computers in their work at least part, if not all, of the time. All in all, this increasing use of technology has helped fuel an unprecedented expansion of the state and national economies. However, along with the benefits, there are several risks for employers. This article will examine some of the basic issues and offer some solutions to business owners who are mindful of the risks involved. First, let's look at some of the risks of the electronic revolution.

Electronic Mail

Electronic mail, or e-mail, has become the communication medium of choice for many employees and businesses. No one doubts its time-saving qualities, but employers must consider the dangers as well:

- Employers can be liable for employees' misuse of company e-mail
- Sexual, racial, and other forms of harassment can be done by e-mail
- Threats of violence via e-mail
- Theft or unauthorized disclosure of company information via e-mail
- E-mail spreads viruses very well

*Public document originally published as Texas Workforce Commission, "Monitoring Employees' Use of Company Computers and the Internet," http://www.twc.state.tx.us/news/efte/monitoring_computers_internet.html

Internet

The Internet is like a super-network connecting countless other computer networks around the world. Literally millions of computers are connected to this vast resource. Every imaginable type of information is available on the Internet if one knows where and how to search for it. As with any kind of resource, it has its good and bad sides. Not surprisingly, employers have had some problems with employees' use of the Internet:

- Unauthorized access into for-pay sites
- Sexual harassment charges from display of pornographic or obscene materials found on some sites
- Trademark and copyright infringement problems from improper use or dissemination of materials owned by an outside party
- Too much time wasted surfing the World Wide Web
- Viruses in downloads of software and other materials from websites

Company Computers

Even with company computers that are not connected to the Internet, employers are finding problems with employees abusing the privilege of having computers to use at work:

- Software piracy—employees making unauthorized copies of company-provided software
- Unauthorized access into company databases
- Use of unauthorized software from home on company computers
- Sabotage of company files and records
- Excessive time spent on computer games
- Employees using company computers to produce materials for their own personal businesses or private use

Many employers wonder what they can do to protect themselves against these kinds of risks and to ensure that company computers and networks are used for their intended purposes. Fortunately, Texas and federal law are both very flexible for companies in that regard. With the right kind of policy, employers have the right to monitor employees' use of e-mail, the Internet, and company computers at work. Doing so successfully requires both a good policy and knowledge of how computers and the Internet work.

Policy Issues

Monitoring employees' use of company computers, e-mail, and the Internet involve the same basic issues as come into play with general searches at work, telephone monitoring, and video surveillance. Those basic issues revolve around letting employees know that as far as work is concerned, they have no expectation of privacy in their use of company premises, facilities, or resources, and they are subject to monitoring at all times. Naturally, reason and common sense supply some understandable limitations, such as no video cameras in employee restrooms, and no forced searches of someone's clothing or body, but beyond that, almost anything is possible in the areas of searches and monitoring. Let's turn to some specifics.

Every employer needs to have a detailed policy regarding use of company computers and resources accessed with computers, such as e-mail, Internet, and the company intranet, if one exists. Each employee must sign the policy—it can be made a condition of continued employment. The policy should cover certain things:

- Define computers, e-mail, Internet, and so on as broadly as possible, with specifics given, but not limited to such specifics
- Define the prohibited actions as broadly as possible, with specifics given, but not limited to such actions
- Remind employees that not only job loss, but also civil liability and criminal prosecution may result from certain actions (illegal pornography, participation in spamming operations or other scams, involvement in computer hacking (see 18 U.S.C. § 1030, among other laws))
- Company needs to reserve the right to monitor all computer usage at all times for compliance with the policy
- Right to inspect an employee's computer, HD, floppy disks, and other media at any time
- Right to withdraw access to computers, Internet, e-mail if needed
- Consider prohibiting camera phones (also called cell phone cameras); such phones have been implicated in gross invasions of other employees' privacy and in theft of company secrets
- Make sure employees know they have no reasonable expectation of privacy in their use of the company's electronic resources, since it is all company property and to be used only for job-related purposes

How to Monitor Compliance

Here is where you as an employer must know at least a few things about computers and the Internet. Naturally, you will leave many of the technical details to certain trusted computer experts on your staff, or you can contract with one of any number of private computer services companies out there. However, you should be armed with some technical knowledge so that you can make better use of the experts' time and be able to tell whether your efforts are successful.

Have your information technology department or computer person set up software monitoring capabilities. Some software can only detect which computer was used on a network, not who used it. An alternative would be to set up a "proxy server"–users have to log in with their own user names and passwords. With regard to the Internet, specific sites can be blocked by Web site addresses and keywords. Some software can analyze the hard drive of each computer on a network, thus establishing who might have unauthorized software or files on their computer.

Where to look for unauthorized computer and Internet activity? On PCs, look in C:\Windows\ for the following folders:

- Cookies—contains "cookies" left on the employee's computer during visits to Web sites—cookies are little files that let Web sites know whether someone has visited the site before
- History—this records the name and Web address of every site visited by the employee
- Temporary Internet Files—this folder contains a copy of every Web page, graphic image, button, and script file found in or on each Web page visited by the employee
- Start Menu: "Documents"–this shows what is in the user's "Recent" folder (recently-opened or recently-used files)

On Macintosh computers, look in the folder for the ISP (Internet Service Provider), then in the folder for the Web browser, then in either "Cache f" or the above names, depending upon what browser the employee uses. The "Apple" menu on Macs also has a "Recent" folder that shows what files the employee has worked on most recently.

With the files found in the above folders, it is possible to reconstruct an employee's entire Web surfing session.

Other places on the computer may yield clues. On PCs, look in the "Recycle Bin"—some people forget to empty that folder when they delete files. Using whatever graphics application you find on the computer, click "File" and look at the recent files in use—you may be surprised at what images the employee has viewed. On Macs, look under "Recent Documents" or double-click the "Trash" icon to see deleted files.

There are some warning signs for computer abuse:

- the employee spends a lot of time online, more than is reasonably needed for the job, yet is strangely non-productive
- you hear a lot of hurried clicking as you approach, and the employee greets you with a red face
- the Temporary Internet Files folder is filled to capacity
- the employee's computer crashes more than anyone else's—viruses and excessive demands on RAM
- an increase in spam e-mail from employees leaving their addresses all over the Internet ("spam" is unsolicited commercial e-mail).

Why Companies Should Be Concerned

Abuse of company computers, networks, and the Internet can leave a company at real risk for an employee's wrongful actions. If an employment claim or lawsuit is filed, it is standard for plaintiff's lawyers and administrative agencies to ask to inspect computer records. Deleting computer files does not completely erase the files—there are many traces left on the user's computer, and forensic computer experts can easily find such traces and use them against a company. Tools exist to make data unretrievable, but not many people are aware of such tools or of how to use them.

An employee in a large semiconductor manufacturing firm was arrested several years ago on charges relating to illicit photos of children after a coworker alerted company managers and the managers called law enforcement authorities. Upon detailed inspection, his office computer was found to have hundreds of illegal images stored on the hard drive. The company's quick action probably prevented what could have been legal problems for the employer itself. In a Central Texas county, a sheriff's department employee was fired after many sexually explicit images were found on his office computer. The department had no problem searching his computer, since it had a well-written policy regarding computer and Internet usage.

The expectation of privacy in workplace electronic systems is important even in the criminal justice context. In the case of *U.S. v. Ziegler*, 474 F.3d 1184 (9th Cir. 2007), *en banc rehearing denied*, 497 F.3d 890 (2007), the Ninth Circuit Court of Appeals found that despite an expectation of privacy in work computers (absent a clear policy to the contrary), the employer can give consent to official searches of such computers, so illegal images of children found on an employee's office computer are admissible as evidence in a criminal case. In a very similar case, *U.S. v. Barrows*, 481 F.3d 1246 (10th Cir. 2007), the Tenth Circuit held that the same result applies, even if the computer is the personal property of the defendant, if the defendant brought the computer to work and took no steps to shield its contents from public inspection (important facts: the defendant used the personal laptop for his work and connected it to the employer's network).

Focus on E-Mail

A good e-mail policy will let employees know that the company's e-mail system is to be used for business purposes only and that any illegal, harassing, or other unwelcome use of e-mail can result in severe disciplinary action. Let employees know that monitoring will be done for whatever purposes. If unauthorized personal use is detected, note the incident and handle it as any other policy violation would be handled. Whatever you do, do not allow employees' personal e-mail to be circulated at random by curious or nosy employees. Such a practice could potentially lead to defamation and invasion of privacy lawsuits. Have your computer experts attach a disclaimer to all outgoing company e-mail that warns of the company's monitoring policy, lets possible unintended recipients know that confidential company information might be included, and disavows liability for individual misuse or non-official use of e-mail. Here is an example of such a disclaimer:

Important Message

Internet communications are not secure, and therefore ABC Company does not accept legal responsibility for the contents of this message. However, ABC Company reserves the right to monitor the transmission of this message and to take corrective action against any misuse or abuse of its e-mail system or other components of its network.

The information contained in this e-mail is confidential and may be legally privileged. It is intended solely for the addressee. If you are not the intended recipient, any disclosure, copying, distribution, or any action or act of forbearance taken in reliance on it, is prohibited and may be unlawful. Any views expressed in this e-mail are those of the individual sender, except where the sender specifically states them to be the views of ABC Company or of any of its affiliates or subsidiaries.

Note: Recent NLRB rulings, employees have the right to use company e-mail systems during non-duty times to discuss with coworkers their terms and conditions of employment.

Court Action

A significant court case in the area of e-mail is *McLaren v. Microsoft Corp.* (No. 05–97–00824-CV, 1999 WL 339015 (Tex.App.—Dallas 1999, no pet)), in which a state appeals court in Dallas ruled that an employee had no claim for invasion of privacy due to the employer's review and distribution of the employee's e-mail. The court noted that having a password does not create reasonable expectation of privacy for an employee, and that since the e-mail system belonged to the company and was there to help the employee do his job, the e-mail messages were not employee's personal property. In addition, the court observed that the employee should not have been surprised that the company would look at the e-mail messages, since he had already told the employer that some of his e-mails were relevant to a pending investigation.

Another court ruled in 2001 that an employer did not violate the federal law known as the Electronic Communications Privacy Act of 1986 (amended by the USA Patriot Act in 2001) when it retrieved an employee's e-mail sent on a company computer to a competitor company in order to encourage the competitor to go after the employer's customers (*Fraser v. Nationwide Mutual Insurance Co.*, 135 F. Supp. 2d 623 (E.D. Pa. 2001)). The employee had sent the e-mail, the recipient at the competitor company had received it, and so the employer had not intercepted the e-mail while it was being sent, which is the only thing protected by the ECPA. On December 10, 2003, the Third Circuit Court of Appeals affirmed that part of the federal district court's judgment (352 F.3d 107).

The New Jersey Supreme Court issued a decision in March 2010 illustrating how important the company's e-mail policy is in determining whether an employee has a reasonable expectation of privacy in e-mail communications and whether an employer steps over the line when reading or monitoring such communications. In *Stengart v. Loving Care Agency*, 990 A.2d 650 (New Jersey 2010), the ex-employee had used a company laptop to communicate with her attorney via a web-based e-mail system in which she had a personal, password-protected account; she did not store the password on the computer. After she left the company, the employer hired a computer forensics expert to make a mirror image of the hard drive. Inspection of the hard drive revealed the e-mails, which the company and its attorney read and used in the course of responding to the employee's lawsuit, even though they were clearly communications between the ex-employee and her attorney, and the e-mails included a standard disclaimer about unauthorized recipients being obligated to destroy the communication, not review it, and notify the sender of the error. The company had a fairly broad computer use policy, but did not define what types of e-mails might be covered, allowed "occasional" personal use of company computers without a notice that any such use would be subject to monitoring, and did not warn employees that information sent, received, or viewed on the computer is stored on the hard drive by the computer's software. Based upon the policy's ambiguity, and on the importance of upholding the principle of attorney-client privilege, the Court ruled that the company's action was an invasion of the employee's privacy and that the company's attorney could potentially be subject to discipline under rules regarding attorney conduct. For a similar case, see *Pure Power Boot Camp, Inc. et al, v. Warrior Fitness Boot Camp, L.L.C., et al.*, 759 F.Supp.2d 417 (S.D.N.Y. 2010).

An important note here: an employer can do anything with e-mail messages sent and received on company computers, even including intercepting them during the process of transmitting or receiving, as long as it has notified employees that they have no expectation of privacy in the use of the company's computer, e-mail, and Internet systems, that all use of such systems may be monitored at any time with or without notice, and that any and all messages, files, and other information sent, relayed, or received with the company's computer, e-mail, and Internet systems are the property of the company, are stored on one or more company computers, and may be subject to company review at any time. All employees may be required to sign a policy acknowledging that they have no expectation of privacy in anything they do on work computers and authorizing the employer to monitor, view, intercept, inspect, copy, store, and further distribute any transmissions that employees send or receive using company electronic equipment or Internet access.

Evidence of Misconduct

If an employee is disciplined or discharged based upon computer or Internet problems, have your company computer experts collect both digital and printed copies of whatever e-mail messages or computer files contain evidence of the violations. The evidence can then be used to defend against various kinds of administrative claims and lawsuits, such as an unemployment claim or discrimination lawsuit.

Conclusion

For business owners, technology makes things both easier and harder. Every company has to ensure that its electronic resources are used properly and not abused by employees. The more that you as an employer know about computers and the Internet, the better off, and safer, your company will be.

41. Disaster Recovery for Technology*

Best Practice

GOVERNMENT FINANCE OFFICERS ASSOCIATION

Governments provide many essential services to their citizens. The disruption of these services following a disaster could result in significant harm or inconvenience to those whom a government serves. State and local governments have a duty to ensure that disruptions in the provision of essential services are minimized following a disaster. Today the public sector, like the private sector, relies heavily upon computers and other advanced technologies to conduct its operations. Therefore, disaster recovery planning, in order to be effective, must specifically address policies and procedures for minimizing the disruption of government operations if computers or other advanced technologies are disabled following a disaster.

Recommendation

GFOA recommends that every government formally establish written policies and procedures for minimizing disruptions resulting from failures in computers or other advanced technologies following a disaster. These written policies and procedures should be evaluated annually and updated periodically, no less than once every three years.

---

*Originally published as Government Finance Officers Association, "Disaster Recovery for Technology: Best Practice," http://www.gfoa.org/disaster-recovery-technology. Reprinted with permission of the publisher.

At a minimum, a government's policies and procedures for computer disaster recovery should do all of the following:

- Formally assign disaster recovery coordinators for each agency or department to form a disaster recovery team. The responsibilities of team members should be defined and a current list of team members and their telephone numbers should be maintained. The government should also establish procedures for assembling the team in the event of a disaster.
- Require the creation and preservation of back-up data. A government's procedures in this regard should cover the regular and timely back-up of computer data (with proper documentation) and the transportation and storage of back-up data off-site (with proper documentation). The government should also ensure the security of back-up data both during transport off site and during storage off site.
- Make provisions for the alternative processing of data following a disaster. A government should enter into a contract for the alternative processing of data following a disaster. It is essential that the government carefully monitor software upgrades to ensure that any such alternative processing site remains capable of processing the government's data. A government should also establish processing priorities should the use of the alternative processing site become necessary. In addition, in situations qualifying for federal emergency assistance, it is essential that the government be capable of providing information to the federal government in the format mandated by the Federal Emergency Management Agency.
- Provide detailed instructions for restoring disk files.
- Establish guidelines for the immediate aftermath of a disaster. Specifically, the government's computer disaster recovery plan should provide guidelines for declaring a disaster, for issuing press releases and dealing with the media, for recovering communications networks, and for assessing damage:
    - A copy of the government's formal computer disaster recovery policies and procedures should be kept off-site to ensure its availability in the event of a disaster;
    - Every government should annually test its computer disaster recovery plan, including communication within the disaster recovery team, and take immediate action to remedy deficiencies identified by that testing. It is essential that such testing encompass the restoration as well as the processing of the government's data; and

- o A government also should satisfy itself concerning the adequacy of disaster recovery plans for outsourced services.

Committee: Accounting, Auditing, and Financial Reporting

NOTES

This best practice was previously titled Planning for Recovery from a Technology Disaster.

Approved by GFOA's Executive Board: March 2007

42. Cybersecurity Partnerships*

Strength in Numbers

SUSAN MILLER

As the public sector wrestles with improving cybersecurity, some organizations are pooling their strengths and forming partnerships to better share threat information and provide tactical cybersecurity training to IT staff.

In North Carolina, the Department of Public Safety is partnering with the Department of Information Technology to form the Information Sharing and Analysis Center. Housed in the state's Bureau of Investigation, ISSAC will promote cyber awareness and information sharing, providing actionable cyber intelligence to private- and public-sector partners and citizens.

ISSAC will work with a number of federal, state and local partners including the North Carolina National Guard, Department of Homeland Security, FBI, U.S. Secret Service, Multi-State Information Sharing and Analysis Center, the State Bureau of Investigation and others.

Recent ransomware attacks like the one in Mecklenburg County in December 2017 and another in Davidson County this February highlight the need for a coordinated response to such threats, state officials said.

"This effort will help us to better guard against cyber threats and to increase information sharing of threat vectors and cyber actor actions across multiple state entities and boundaries," N.C. Department of Information Technology Secretary Eric Boyette said. "With the increased coordination and sharing of information will come an increase in the speed with which we can detect, identify and recover from cyber incidents."

---

*Originally published as Susan Miller, "Cybersecurity Partnerships: Strength in Numbers," *GCN*, Mar 22, 2018. Reprinted with permission of the publisher.

The Secure Campus Enterprise

Cybersecurity information sharing across university enterprise networks will be easier with the launch of *OmniSOC*, a cybersecurity operations center that will provide real-time intelligence sharing and threat analysis for its five university members. A joint initiative of Indiana University, Northwestern University, Purdue University, Rutgers University and the University of Nebraska–Lincoln, OmniSOC's goal is "to help higher education institutions reduce the time from first awareness of a cyber security threat anywhere to mitigation everywhere for members," according to a news announcement.

Operating out of Indiana University, OmniSOC combines real-time security data feeds from its member campuses with governmental and corporate security subscriptions, and uses that information to identify suspicious and malicious activity, officials said. It then provides rapid incident response based on both human analysis and machine learning.

OmniSOC uses the Elastic Stack security analytics platform, a system for ingesting, correlating and analyzing vast quantities of information to detect cyber threats.

"With tens of thousands of students, faculty and staff, university campuses are really like small cities, with sensitive data and powerful computing systems that are coveted by cyber criminals," Tom Davis, OmniSOC founding executive director and chief information security officer, said in a statement. "While campus-by-campus approaches are essential, they are not sufficient for the sophistication of modern cyber risks."

"Higher education is for the most part an open environment, so we often see cyber crimes that others have not," Purdue University Chief Information Security Officer Greg Hedrick said. "By allowing us to monitor across higher education, OmniSOC helps to improve our capabilities to identify and react more quickly to these bad actors. My hope is that this information can be shared with others outside of our community in order to protect the entire ecosystem."

OmniSOC plans to scale up services and expand its membership to other universities.

Better Cyber Awareness

To tackle training and education, the University of West Florida's Center for Cybersecurity is partnering with the Florida Agency for State Technology to better prepare state personnel to detect and protect against emerging cyber threats. The program will offer hands-on training and educational courses using face-to-face, online and remote delivery, and will provide competency-based certifications to prepare state personnel for core cybersecurity work roles.

Initial training for IT staff began March 20 at the Florida Department of Revenue, focusing on cybersecurity awareness and fundamentals and will eventually cover cybersecurity incident management, network defense, operating system hardening, risk management, cloud security and other emerging topics.

The Florida Cyber Range and UWF Cybersecurity for All program will be used to provide training. The cyber range is a high-fidelity training environment can emulate the internet, replicate websites, integrate social media and support dynamic interjection of vulnerabilities. The program provides training on emerging topics via an online learning environment, customizable modules and hands-on activities using the cyber range.

"As the threats evolve, we must continue to train our information security and technology resources," AST Executive Director and State CIO Eric Larson said. "We hope our partnership with UWF will serve as a model for other states to not only provide advanced cyber training, but to offer ongoing educational opportunities for state employees."

Rhea Kelly, executive editor at Campus Technology, a sibling site to GCN, contributed to this story.

43. Intersector Briefing*

Cross Sector Collaboration in Cybersecurity

INTERSECTOR PROJECT

In 2016, it seemed that cybersecurity, privacy and hacking were constantly in the news, from the heated discussion of election related hacking to the revelation that in December 2013, Yahoo had been attacked, compromising the personal information of hundreds of millions of consumers. The vulnerability of private information to cyber threats is an issue that spans across the private and public sectors, affecting not only our federal, state and local governments, but also small businesses and large multinational enterprises and individuals across the country.

Cybersecurity is a complex issue that not only affects cross sector stakeholders, but also must involve cross sector collaboration to make the United States less vulnerable to cyber attacks. This Intersector Briefing looks at public-private collaboration for cybersecurity, from improving communication and information between sectors at all levels of government to stop attacks, to building partnerships to improve cybersecurity education.

Lawmakers Receive Lukewarm Assessment of Cyber Cooperation Between Feds, Private Sector

This piece from *The Hill* discusses coordination between the public and private sectors around cybersecurity, after a congressional hearing earlier

*Originally published as The Intersector Project, "Intersector Briefing: Cross Sector Collaboration in Cybersecurity," *PA Times*, April 25, 2017. Reprinted with permission of the publisher.

this month. The overall message of industry experts was that the Department of Homeland Security (DHS) needs to share more information more quickly with private organizations around cyber threats.

"Our collective ability to combat these threats, with government and the private sector working together, will be one of the defining public policy challenges of our generation," said Rep. John Ratcliffe (R–Texas), chair of the Homeland Security Committee's Cybersecurity Subcommittee, in opening testimony. The hearing precedes the release of President Trump's executive order on cybersecurity, with industry leaders taking the hearing as an opportunity to share their thoughts on what they'd like to see happen, namely the new administration "improving the implementation of partnerships between the DHS and private companies."

Amazon Partners with State for New Cybersecurity Education Pipeline

While collaboration between the public and private sectors to address direct cyber threats is important, there are other ways that cross sector collaboration can help strengthen cybersecurity in the United States. For instance, there is a "critical cybersecurity expert shortage," explains cybersecurity expert Algirde Pipikaite in an opinion piece in *The Hill*. Much like other workforce development or STEM education programs, there is a role to play for both public and private sectors.

In one attempt to improve cybersecurity education, Amazon Web Services is teaming up with Virginia Cyber Range, a Commonwealth of Virginia initiative, to enhance cybersecurity education in the state's high schools, colleges, and universities. "In Virginia and across the country, businesses, governments, and private individuals are impacted by the growing threat of cyber-attacks," explained Virginia Governor Terry McAuliffe. "We need a capable workforce that understands these swiftly changing threats and is ready to mount an agile defense against them."

Cybersecurity Partnerships: A New Era of Public-Private Collaboration

This 2014 report from the Center on Law and Security at the NYU School of Law acknowledges the need for public-private collaboration for cybersecurity, but recognizes its complexities. "Legal, strategic, and pragmatic obstacles often impede effective public-private sector cooperation, which are compounded by regulatory and civil liability risks. Different government agencies have competing roles and interests, with the government serving dual roles as both partner and enforcer, influencing how companies facing cyberthreats view public authority," the author writes. The report goes on to look at why public-private collaboration is valuable to addressing cybersecurity challenges, the barriers that impede it, the best methods of cross-sector collaboration for these efforts and more.

Challenges Ahead for New White House Cybersecurity Advisor

President Trump is poised to select Rob Joyce, currently Chief of the National Security Agency's Tailored Access Operations, as his cybersecurity czar, said a *Forbes* article in March. Since Edward Snowden's reveal of the "insidious inner-workings and questionable ethics of the NSA, there has been lingering concerns over privacy and trust between that organization and private industry and citizens," Tony Bradley reports, noting the potential for that distrust to reverberate through this administration's cybersecurity work, with Joyce at the helm.

This could be detrimental at a time when the cybersecurity sector needs increased collaboration between the sectors. "The U.S. Cyber Czar's role is more important now than ever before," says Fleming Shi, SVP of Advanced Engineering at Barracuda. "It's … critical to our national security to create a true partnership between private industry and the government. Not just consulting and regulating, but real collaboration."

San Diego Cybersecurity Chief Shares Three Ways to Shield Cities from Attacks

In an interview with *State Scoop*, San Diego's Chief Information Security Officer Gary Hayslip shares strategies for improving cybersecurity at a city level. After working in the Defense Department, Hayslip had to adjust to the realities of working in a city, realizing that "the City is really a $4 billion business with 1.3 million customers and many departments. … It requires a different approach." He notes the importance of partnerships between City departments and the private sector, with one result of his work in San Diego being increased coordination between the City's cybersecurity teams and startups to "share the same vision for a cybersecurity roadmap that protects citizen data."

44. Obama's Cybersecurity Initiative*

Frank J. Cilluffo *and* Sharon L. Cardash

The linchpin of President Obama's recently launched cybersecurity initiative is to encourage the private sector to share information to better defend against cyberattacks.

Yet U.S. companies have historically been wary of openly talking about their cybersecurity efforts with competitors and with government—for good reason.

Many businesses fear that sharing threat-related information could expose them to liability and litigation, undermine shareholder or consumer confidence, or introduce the potential for leaks of proprietary information.

For some companies, Edward Snowden's revelations of sweeping government surveillance programs have reinforced the impulse to hold corporate cards close to the vest. Yet on the heels of a deluge of high-profile cyberattacks and breaches against numerous U.S. companies, we may finally have reached a tipping point, where potential harm to reputation and revenue now outweighs the downside of disclosure from a corporate perspective.

Blueprint for Safer Internet

Obama's executive order is meant to shore up public health and safety, as well as national and economic security, by promoting the exchange of information on cybersecurity risks and incidents. The goal is to share data

*Originally published as Frank J. Cilluffo and Sharon L. Cardash, "Obama's Cybersecurity Initiative," *The Conversation*, February 24, 2015. Reprinted with permission of the publisher.

within and between industries to foster speedy and effective response to cyberthreats.

The executive order empowers the Secretary of Homeland Security to "strongly encourage the development and formation of Information Sharing and Analysis Organizations" (ISAOs), "organized on the basis of sector, subsector, region, or any other affinity, including in response to particular emerging threats or vulnerabilities." These ISAOs are intended "to serve as focal points for cybersecurity information sharing and collaboration within the private sector and between the private sector and government."

In addition, three days before the announcement of the executive order, the White House announced the creation of a national Cyber Threat Intelligence Integration Center (CTIIC). Akin to the National Counterterrorism Center, the CTIIC will work to "connect the dots between various cyberthreats to the nation so that relevant departments and agencies are aware of these threats in as close to real time as possible." The ultimate objective is to "facilitate and support efforts by the government to counter foreign cyberthreats."

The idea underlying the executive order and companion measures is to make it harder for cybercriminals and worse to achieve their prize, be it profit, intellectual property, state secrets, or geo-strategic advantage. For too long, too many factors have operated in the cyberattacker's favor.

Despite the fact that the executive order lacks the force of legislation (only Congress can introduce liability protections, for instance), private sector companies may choose to cooperate. And corporate compliance, while voluntary, is crucial because more than 80% of U.S. critical infrastructure is privately owned and operated. Each such sector is, in and of itself, essential to U.S. national and economic security.

Models for Cooperation

Keeping enterprises up and running is all the more important because their operations may be intertwined with one another. Taking down one sector, such as the electric grid for example, may therefore bring down others, yielding cascading and potentially catastrophic effects for the country. The good news is that collaboration between and among private entities is already underway, and one size need not fit all.

Take, for example, the Financial Services Information Sharing and Analysis Center (FS-ISAC), which facilitates sector-wide exchanges regarding cyber-related threats and their remediation. Or consider Microsoft's Cybercrime Center, which works in tandem with law enforcement and other partners worldwide to disseminate information and thwart criminals. These are just two examples of corporate actors spearheading initiatives that pre-date the executive order and that serve both the public and private interest.

Letting a thousand flowers bloom—or encouraging flows of information between industries and government—may seem like a chaotic approach, yet existing efforts have achieved some real success. More such endeavors, tailored to context, may in fact prove constructive as the cyber-threat ecosystem continues to evolve.

For example, a group of U.S. companies (including McAfee and Symantec) are banding together to form a "Cyber Threat Alliance" which aims "to disperse threat intelligence on advanced adversaries across all member organizations to raise the overall level of situational awareness to better protect both the ... organizations and their customers." After all, it is companies themselves that usually have the greatest incentives to protect their own assets. Yet companies need to understand and respect the contours of what constitutes lawful defense and response, consistent with government's rules of the road which, admittedly, are a work in progress, at best.

Other countries are also grappling with the question of how to effectively protect systems and networks, both private and public. Leading the pack is Estonia, an early target of cyberattack (2007) and an early adopter of e-governance (government services provided online), with a continuing commitment to innovation and digital security that is widely shared by officials and entrepreneurs alike. The country's latest cyber-initiative is bold and ambitious: creating "digital data embassies" worldwide and offering "digital citizenship" ("e-residency rights") to all who are prepared to meet the requirements. This gambit has dual goals: protect data and services in the event of cyber-attack and, secondly, facilitate additional foreign investment in the country and thereby generate economic growth.

National Imperative and Individual Duty

What works for Estonia may not be a good fit—at least in totality—for other nations. The country is small in terms of terrain and population, and did not have to contend with legacy issues when building its infrastructure from the ground up after regaining their independence from Soviet rule in 1991. But the principles of Estonia's policies are certainly instructive.

These include a whole-of-society approach to cybersecurity that incorporates the discipline (coding, programming, etc) into the education system and curricula, beginning in first grade and continuing through to university. The result is a prevailing culture and mindset that conceives of cybersecurity as both a national imperative and an individual duty.

As the United States seeks to elevate its cybersecurity posture in ways that best suit its size, economy, circumstances, and traditions (based on history, respect for privacy and civil liberties, and so on), it will be important to complement private sector information-sharing efforts with a host of other measures.

These include building a cyberworkforce that is sufficiently large and skilled to meet existing and future U.S. needs. It means designing and engineering secure systems and architectures. It also includes cultivating an operating culture (in government and business) that recognizes cybersecurity to be a priority from the get-go as opposed to an afterthought. Falling short here will negatively affect U.S. national and economic security.

This month's executive order is a spur to get the ball rolling but, frankly, there is a limit to what government alone can (and should) do in this area. Changes in attitudes and behaviors are needed across the board, right down to families and individuals.

45. New Trump Executive Order on Cybersecurity*

DANIEL J. LOHRMANN

After campaign promises on cyber, months of tough talk about Internet security plans, plenty of anticipation and a missed 90-day deadline to deliver a cybersecurity report, President Donald Trump signed an Executive Order (EO) on cybersecurity this week.

The Presidential Executive Order on Strengthening the Cybersecurity of Federal Networks and Critical Infrastructure offers three sections, which Tom Bossert, Trump's homeland security adviser, said were in priority order:

Section 1. Cybersecurity of Federal Networks
Section 2. Cybersecurity of Critical Infrastructure
Section 3. Cybersecurity for the Nation

News media and overall cyberindustry reaction to the EO have been mostly positive.

Cyber Executive Order Details

The EO starts with the clear policy that: "The President will hold heads of executive departments and agencies (agency heads) accountable for managing cybersecurity risk to their enterprises."

Next, the findings, which outline inadequate cyberdefenses in federal agencies, also make it clear that the status quo will not be tolerated. An exam-

*Originally published as Daniel J. Lohrmann, "New Trump Executive Order on Cybersecurity," *Government Technology*, May 14, 2017. Reprinted with permission of the publisher.

ple: "The executive branch has for too long accepted antiquated and difficult-to-defend IT." Also, "Effective risk management requires agency heads to lead integrated teams of senior executives with expertise in IT, security, budgeting, acquisition, law, privacy, and human resources."

More specifically, "Effective immediately, each agency head shall use The Framework for Improving Critical Infrastructure Cybersecurity (the Framework) developed by the National Institute of Standards and Technology, or any successor document, to manage the agency's cybersecurity risk. Each agency head shall provide a risk management report to the Secretary of Homeland Security and the Director of the Office of Management and Budget (OMB) within 90 days of the date of this order."

While a few experts in the field, including former White House cybersecurity coordinator Michael Daniel, called this EO just "A plan for a plan," these directives will be difficult risk management reports for agencies to complete in three months.

The section on critical infrastructure builds on what was done during the Obama administration. The EO starts with this policy: "It is the policy of the executive branch to use its authorities and capabilities to support the cybersecurity risk management efforts of the owners and operators of the Nation's critical infrastructure (as defined in section 5195c(e) of title 42, United States Code) (critical infrastructure entities), as appropriate."

The president goes on to outline how that protection effort will be done and who will be involved.

Another report is due in 90 days regarding "appropriate market transparency of cybersecurity risk management practices by critical infrastructure entities."

Within 240 days, a report is due on our "resilience against botnets and other automated, distributed threats."

The electric grid is specifically called out with an "assessment of electricity disruption incident response capabilities." (That is, are we ready for an attack against the electric grid?) This report is due in 90 days as well.

Finally, another report due in 90-days will address cybersecurity risks facing the defense industrial base, including its supply chain, and United States military platforms, systems, networks and capabilities, and recommendations for mitigating these risks.

In the area of "cybersecurity of the nation," the policy reiterates our priorities that "open, interoperable, reliable, and secure internet that fosters efficiency, innovation, communication, and economic prosperity, while respecting privacy and guarding against disruption, fraud, and theft." There is also the goal of fostering a next-generation workforce that is skilled in cybersecurity.

My Viewpoint

In many ways, this EO lays out the critical agenda for high-priority action items in cyberspace for the next four years. It offers a mix of different themes and topics that is diverse, from critical infrastructure to a cyberworkforce.

I view this as just the beginning for the Trump administration plans for cyberspace. While some may say that the words and deeds prior to this were actually the opening act, most of those statements were not backed up with an executive order with guidance to various groups to get moving.

These reports and other deliverables will be essential building blocks with much more to come. This is a foundational EO on cyber that continues the momentum that was built in the Obama administration, but also adds much more federal agency director accountability. This is a good thing, since every cyberexpert knows that true management buy-in and support is a critical success factor.

I am hearing that that there is also more going on behind the scenes right now that this EO reveals. For example, Rudy Giuliani is helping draw up cyber doctrine, DNI says, but details are scarce. I also think the international cooperation piece of this cybersecurity EO is essential. The EO directs:

> Within 45 days of the date of this order, the Secretary of State, the Secretary of the Treasury, the Secretary of Defense, the Secretary of Commerce, and the Secretary of Homeland Security, in coordination with the Attorney General and the Director of the Federal Bureau of Investigation, shall submit reports to the President on their international cybersecurity priorities, including those concerning investigation, attribution, cyber threat information sharing, response, capacity building, and cooperation. Within 90 days of the submission of the reports, and in coordination with the agency heads listed in this subsection, and any other agency heads as appropriate, the Secretary of State shall provide a report to the President, through the Assistant to the President for Homeland Security and Counterterrorism, documenting an engagement strategy for international cooperation in cybersecurity.

Final Thoughts

The importance of this cyberdefense topic was underlined on Friday, when a new global ransomware attack called WannaCry was unleashed that affected over 100 countries and shut down many hospitals and businesses worldwide. This ongoing situation is one of the largest cyberattacks ever.

It was almost as if the response to the president's cybersecurity EO from global hackers was, "Our life goes on and we don't really care what you do." This is our sad, but scary, online reality.

We all need to be reminded that our individual and corporate (cybersecurity industry) actions have a great ability to influence lives all over the planet—both online and offline. A renewed urgency is required in cyberspace, as our online problems are not going away.

The second chapter in Trump's cybersecurity plan will begin when those reports and actions steps are due later this year. Meanwhile, our cyberbattles march on.

46. Sharing of Cyber Threat Indicators and Defensive Measures by the Federal Government*

OFFICE OF THE DIRECTOR OF NATIONAL INTELLIGENCE AND OTHER AGENCIES

Section 103 of the Cybersecurity Information Sharing Act of 2015, Pub. L. 114–113, 129 Stat.694 (2015), directs the Director of National Intelligence, the Secretary of Homeland Security, the Secretary of Defense, and the Attorney General, in consultation with the heads of the appropriate federal entities set forth in Subsection 1.1, to jointly develop and issue procedures to facilitate and promote:

1. Timely sharing of classified cyber threat indicators (CTIs) and defensive measures (DMs) in the possession of the Federal Government with representatives of relevant federal entities and non- federal entities that have appropriate security clearances;
2. Timely sharing with relevant federal entities and non-federal entities of cyber threat indicators, defensive measures, and information relating to cybersecurity threats or authorized uses under this title, in the possession of the Federal Government that may be declassified and shared at an unclassified level;

---

*Public document originally published as Office of the Director of National Intelligence, Department of Homeland Security, Department of Defense, and Department of Justice, "Sharing of Cyber Threat Indicators and Defensive Measures by the Federal Government," (February 16, 2016)

3. Timely sharing with relevant federal entities and non-federal entities, or the public if appropriate, of unclassified, including controlled unclassified, cyber threat indicators and defensive measures in the possession of the Federal Government;

4. Timely sharing with federal entities and non-federal entities, if appropriate, of information

5. relating to cybersecurity threats or authorized uses under this title, in the possession of the Federal

6. Government about cybersecurity threats to such entities to prevent or mitigate adverse effects from such cybersecurity threats; and

7. Periodic sharing, through publication and targeted outreach, of cybersecurity best practices that are developed based on ongoing analyses of cyber threat indicators, defensive measures, and information relating to cybersecurity threats or authorized uses under this title, in the possession of the Federal Government, with attention to accessibility and implementation challenges faced by small business concerns (as defined in Section 3 of the Small Business Act (15 U.S.C. 632)).

The procedures outlined in this document describe the current mechanisms through which the appropriate federal entities, as named in Section 102(3), share information with non-federal entities.

Examples of non-federal entities are private sector entities and state, local, tribal and territorial (SLTT) governments, including owners and operators of private and public critical infrastructure. These procedures are implemented today through a series of programs, which are described below and provide the foundation of appropriate federal entities' cybersecurity information sharing capability. These programs are dynamic and are expected to grow or evolve over time. That said, some programs may be discontinued and new programs may begin. In addition, these programs work together to identify useful information available through their unique information sources and to share that information with their respective partners. Wherever possible, appropriate federal entities coordinate with each other through these programs to ensure that the information they share is timely, actionable, and unique.

Federal entities are encouraged to share CTIs and DMs as broadly and as quickly as possible. Whether CTIs and DMs are classified, declassified or unclassified, federal entities should continuously identify and implement programs to share such CTIs and DMs with each other and with non-federal entities.

Federal entities engaging in activities authorized by CISA, including those referenced within this document, shall do so in full compliance with the Constitution and all other applicable laws of the United States, Executive Orders, and other Executive Branch directives, regulations, policies and procedures, court orders and all other legal, policy and oversight requirements.

In furtherance of this general encouragement to share broadly and quickly, federal entities shall establish and maintain procedures; and consistent with those procedures, maintain programs that:

1. Facilitate the timely sharing of classified CTIs and DMs in the possession of the Federal Government with representatives of relevant federal entities and non-federal entities that have appropriate security clearances.
2. Share with other relevant federal entities and non-federal entities CTIs, DMs, and information relating to cybersecurity threats in their possession that may be declassified and shared at an unclassified level. Such sharing is consistent with the emphasis placed by the President and the Director of National Intelligence on the need to ensure the timely and efficient flow of CTIs and DMs to appropriate federal and non-federal entities and shall be conducted consistent with all applicable Executive Orders and directives.
3. Support the timely sharing with relevant federal entities and non-federal entities, or the public if appropriate, of unclassified, including controlled unclassified, CTIs and DMs in the possession of the Federal Government.
4. Support the timely sharing with federal entities and non-federal entities, if appropriate, of information relating to cybersecurity threats or authorized uses under CISA, in the possession of the Federal Government about cybersecurity threats to such entities to prevent or mitigate adverse effects from such cybersecurity threats.
5. Support the periodic sharing, through publication and targeted outreach, of cybersecurity best practices that are developed based on ongoing analyses of CTIs, DMs, and information relating to cybersecurity threats or authorized uses under this title, in the possession of the Federal Government, with attention to accessibility and implementation challenges faced by small business concerns.

This document sets forth relevant procedures, or otherwise references exemplar activities that have implemented such procedures. In addition, this document provides that federal entities will share with each other as a means of also sharing more broadly with non-federal entities since many federal entities maintain unique relationships with different cross-sections of the Nation, such as critical infrastructure sectors, regulated industries or State and local governments. Finally, this document recognizes that broad sharing within components of a federal entity can be just as important as broad sharing between federal entities.

It is the policy of the U.S. Government to make every reasonable effort "to ensure the timely production of unclassified reports of cyber threats to the U.S. homeland that identify a specific targeted entity."

Sharing of cyber threat information that is classified, however, is dependent upon the recipient's security clearance level and must be performed in accordance with applicable policy and protection requirements for intelligence sources, methods, operations, and investigations, which are not superseded by this document. Any federal entity sharing classified information must continue to conform to existing classification standards and adhere to handling restrictions, like Originator Controlled (ORCON) markings or specific originator instructions on use of downgraded information, when determining what information can be shared with any entity. Given the protections for and sensitive nature of classified information, additional emphasis must be placed on coordination early in the process, with originators of specific classified information deemed necessary to share with an entity.

When appropriate, agency heads are expected to continue using the emergency authority granted in 32 CFR Section 2001.52, promulgated pursuant to Executive Order 13526—*Classified National Security Information*, to disseminate and transmit classified information during certain emergency situations, in which there is an imminent threat to life or in defense of the homeland, to those who are otherwise not routinely eligible for access.

The following programs are a non-exhaustive set of examples that use current procedures to support the timely sharing of classified CTIs and DMs in the possession of the Federal Government with representatives of relevant federal entities and non-federal entities that have appropriate security clearances.

Department of Homeland Security (DHS) Enhanced Cybersecurity Services (ECS) Program—http://www.dhs.gov/enhanced-cybersecurity-services

The DHS ECS program is a voluntary information sharing program that assists U.S.-based public and private entities as they improve the protection of their computer systems from unauthorized access, exploitation, or data exfiltration. DHS works with cybersecurity organizations from across the Federal Government to gain access to a broad range of sensitive and classified cyber threat information. DHS develops CTIs based on this information and shares them with qualified commercial service providers (CSPs), thus enabling them to better protect their customers. ECS augments, but does not replace, entities' existing cybersecurity capabilities.

The ECS program does not involve government monitoring of private networks or communications. Under the ECS program, information relating to cyber threats and malware activities detected by the CSPs is not directly shared between CSP customers and the Federal Government. However, when a CSP customer voluntarily agrees, the CSP may share limited and anonymized information with DHS.

In February 2013, Executive Order 13636, *Improving Critical Infrastructure Cybersecurity*, expanded ECS to each of the 16 critical infrastructure sectors. As a result of increased demand and need for cybersecurity protection across the nation, the ECS program has since expanded further and now allows approved CSPs to extend their ECS customer base to all U.S.-based public and private entities.

Department of Defense (DoD) Defense Industrial Base (DIB) Cybersecurity (CS) Program—32 CFR Part 236, http://dibnet.dod.mil/

The DIB CS Program was initiated in 2007 and established as a permanent DoD program in 2013 under 32 Code of Federal Regulations, Part 236, to enhance and supplement DIB participants' capabilities to safeguard DoD information that resides on, or transits, DIB unclassified networks or information systems. Under the voluntary DIB CS program, DoD and DIB participants share cyber threat information in order to enhance the overall security of unclassified DIB networks, reduce damage to critical programs, and increase DoD and DIB cyber situational awareness.

The DoD Cyber Crime Center (DC3) serves as the operational focal point for the DIB CS program, sharing cyber threat information with DIB participants in near real-time at both the classified and unclassified levels. Participating companies receive analytic support, incident response, mitigation and remediation strategies, malware analysis, and other cybersecurity best practices.

Information shared between DoD and the DIB under the DIB CS Program strengthens the Nation's knowledge of the ever-growing cyber threat, increases the effectiveness of mitigating the risk, and meets the Administration's and DoD's strategic objective of enhancing voluntary government-private sector cyber threat information sharing.

DHS Cyber Information Sharing and Collaboration Program (CISCP)—http://www.dhs.gov/ciscp

The Cyber Information Sharing and Collaboration Program (CISCP) is DHS's flagship program for public-private information sharing and complement ongoing DHS information sharing efforts. In CISCP, DHS and participating companies share information about cyber threats, incidents, and vulnerabilities. To join CISCP, companies are required to sign a Cooperative Research and Development Agreement (CRADA). Along with governing participation in CISCP, a signed CRADA may permit access to the National Cybersecurity and Communications Integration

Center (NCCIC) watch floor and allows for company personnel to be eligible for security clearances to view classified threat information.

The National Cyber Investigative Joint Task Force (NCIJTF)

The NCIJTF is a Presidentially-mandated multi-agency cyber center that coordinates, integrates, and shares information related to cyber threat investigations and operations. The NCIJTF currently has signed memoranda of understanding (MOUs) with approximately 24 member agency representatives, which allow for sharing of cyber threat information—to include classified CTIs—at the NCIJTF. The appropriate federal entities identified under Section 102(3) are current members of the NCIJTF with signed MOUs.

The NCIJTF has several existing mechanisms for sharing classified CTIs to the appropriate federal entities, as members of the NCIJTF. CyWatch, the NCIJTF's 24/7 watch floor, serves as the primary mechanism for sharing classified CTIs with federal entities that are NCIJTF members. In addition, the NCIJTF's Office of Threat Pursuit analyzes collected cyber threat data and provides reports on exfiltrated data, which are shared with member agencies. Lastly, the Office of Campaign Coordination facilitates the sharing of classified CTIs and DMs related to campaign missions among participating agencies.

The NCIJTF also provides classified threat briefings to both federal entities and non-federal entities, to include cleared private sector representatives. Briefings are determined on an ad-hoc basis.

In addition to sharing through the NCIJTF, the FBI utilizes on-site briefings to share classified indicators and defensive measures with industry and appropriate private sector entities. Coordinating with its other government agency partners, the FBI provides potential or known victim entities with temporary security clearances so they may have access to specific classified information and technical indicators that may be used to neutralize an ongoing threat. Oftentimes, the technical information exchanged is accompanied by a contextual briefing to emphasize the severity of the threat.

47. Law Enforcement Partnerships Enhance Cybersecurity Investigations*

CHELSEA BINNS

Collaboration is an important element of successful police work. Law enforcement routinely works with its citizens, and one another, to fight crime. The proliferation of cybercrime has brought a renewed interest in law enforcement partnerships with the private sector. These collaborations have facilitated the sharing of information and resources to enhance investigations. Recent cases demonstrate their value.

Cybercrime is now one of our biggest threats in society. Cybercrime is expected to cost $6 trillion in damages globally by 2021. This growth is attributed in part to an increase in "organized crime gang hacking activities." Recent schemes demonstrate today's cybercriminals are working in large groups across the globe to perpetuate their crimes. Thus, law enforcement is joining forces with the private sector to augment and extend their investigative power.

One example is The Department of Homeland Security's Information Network (HSIN). The HSIN, created ten years ago, provides a way for law enforcement to "collaborate securely with partners across geographic and jurisdictional boundaries." It now connects 18,000 law enforcement, 60,000 responding agencies and 78 fusion centers. In 2015, the HSIN expanded their network to include "private sector stakeholders" and "infrastructure owners, emergency managers and cybersecurity experts."

Recent cases demonstrate the value of such collaborations. In November

---

*Originally published as Chelsea Binns, "Law Enforcement Partnerships Enhance Cybersecurity Investigations," *PA Times*, April 4, 2017. Reprinted with permission of the publisher and author.

2016, an investigative team solved a major cybercrime resulting in 178 arrests. In this case, 580 "money mules" caused losses of EUR 23 million (approx. $24 million USD). The team, the European Money Mule Action, (EMMA) joined law enforcement from 17 countries and two federal agencies, with 106 banks and private partners. This partnership allowed law enforcement to share data and other intelligence in a coordinated operation. The EMMA has built on their success by adding more partners in the private sector. They have also launched an awareness campaign to prevent future cybercrimes.

Private sector partnerships add critical value to cybercrime investigations. In fact, experts say it's "impossible" to effectively address cybercrime without the two parties working together. The private sector has an important role in the prevention and detection of cybercrime. It owns and operates 85 percent of all critical infrastructures, and often is the "first line of defense" against potential threats. Law enforcement also benefits from the additional crime-fighting resources the private sector provides. This assistance can be especially useful, due to the void of Information Technology-related talent in the government sector.

European law enforcement, which reports a "relentless growth of cybercrime" is capitalizing on private sector resources. They find cybercrime is becoming increasingly professionalized, because of strong financial and operational support from organized crime groups. Comparatively, their law enforcement is "under-resourced." For them, collaborating with the private sector is a "smarter approach" which provides them with the tools they need to ensure a successful investigation.

Law enforcement in New Zealand is currently leveraging resources in academia to combat their cybercrime problem. Specifically, police are collaborating with researchers at Waikato University, to benefit from their expertise in computer science, data mining and software engineering. In 2015, New Zealand was subjected to 8570 cyber-attacks, costing an estimated $13.4 million.

Partnerships also help foster the flow of information. There is an unfortunate history of information gaps between the government and the private sector, which can provide opportunity for cybercriminals. For instance, in 2014, Google located a serious, national security vulnerability, that "allowed attackers to steal encrypted information, including passwords, cookies and data." It was later learned this flaw had been previously discovered by the National Security Agency (NSA) years earlier, yet Google didn't know it, because the two parties did not collaborate.

There are many exciting partnerships currently underway to combat cybercrime threats. The U.S. Attorney's Office's Criminal Division in Colorado recently created a new Cybercrime and National Security Section, which aims to work with private industry towards cybercrime prevention.

The U.S. Department of Homeland Security, the Department of Defense, and the Air Force have all recently established a presence in Silicon Valley, to foster cyber security collaboration with technology firms.

While there are many benefits to these alliances, there are also challenges. Organizational politics, culture and bureaucracy have served as roadblocks to past collaborative efforts.

Silicon Valley and the federal government are one alliance facing such challenges. Their cultures clash. Silicon Valley's organizational structure is lean and quick, while the federal government is comparatively large and slow. Regardless, although keenly aware of the "burden of Federal bureaucracy," Silicon Valley technology firms typically "downplay[] these concerns." They view "government work as an exciting and fulfilling way to defend the U.S. in cyberspace." Their positive experience may inspire other private firms to forge the same battle.

Experts have also suggested offering incentives to the private sector to partner with government. For instance, Retired NSA Director General Keith Alexander suggests providing tax incentives to private companies to develop robust cyber security programs and share their findings with the government.

Overall, incentive or no incentive, it is clear the public and private sectors are forming successful strategic partnerships. It is a worthwhile trend that will hopefully continue to flourish and add value to global cybercrime investigations.

48. States Rush to Cash in on Cybersecurity Boom*

ELAINE S. POVICH

As data dragnets and information breaches dominate the news, states are scrambling to cash in on a rapidly expanding business sector by offering tax incentives to firms that protect sensitive information from outside attacks.

While ordinary Americans wonder if their private phone calls and emails are being monitored by their government, businesses are concerned that proprietary and sensitive business information could be stolen by competitors—at home and from overseas. State and local governments also are working to tighten safeguards to prevent outsiders from hacking into their information.

"It's the new global threat, not only to our state and nation, but to the whole world," said Mark A. Vulcan, program manager at the Maryland Department of Business and Economic Development.

Maryland is breaking new ground with a total $3 million offer of tax breaks to be distributed among cybersecurity startups already in the state or who agree to locate there. While many states include cybersecurity companies in their overall tax incentives for high tech firms, Maryland's legislation—proposed by Gov. Martin O'Malley and signed in May—appears to be unique.

What also sets Maryland apart from other states, Vulcan said, is that this tax credit goes directly to the company, not the "angel" investor in that entity, which many other states do.

Analysts say this credit could signal a new wave of action by states trying to cash in on the cybersecurity boom. The $207 billion cybersecurity industry is expected to show "impressive growth" over the next five years, according to Entrepreneur.com, a site for investors.

---

*Originally published as Elaine S. Povich, "States Rush to Cash in on Cybersecurity Boom," *Governing*, June 17, 2013. Reprinted with permission of the publisher.

Consultant Javier Siervo, with the Berkeley Research Group, LLC, in Washington, D.C., said Maryland may be the only state offering a tax credit specifically for cybersecurity, but D.C. offers incentives to companies that have an office in the District and "derive most of their revenue from technology-related activities." And nearby Arlington County in Virginia has increased technology zones to encourage tech businesses to move operations to the county.

In Virginia, a statewide program that offers capital gains tax exemptions to tech companies would cover cybersecurity companies as well as other high-tech ventures, according to Cameron Kilberg, the state's assistant secretary of technology. Under that incentive, the state doesn't tax any income already taxed by the federal government as a long-term capital gain.

The program began in 2010, Kilberg said, and will continue to 2015, past its original sunset date of 2013, because the state wanted more time to evaluate whether the program was effective.

Cybersecurity Companies

The area around Washington is home to many government contractors and attracts a sizable cybersecurity industry, said Michael Colavito, state and local tax expert at Aronson LLC, who advises private business on how to take advantage of state tax incentives. He said security needs to be stepped up "because of all the hackers out there," and Maryland is trying to position itself to take advantage of that situation.

Some of the nation's largest defense and security companies are among the top 20 worldwide in the cybersecurity business, including Booz Allen Hamilton Inc. (the company of Edward Snowden, who leaked the National Security Agency's data dragnet program), General Dynamics, Intel, Lockheed Martin, Northrop Grumman and Raytheon among others.

Much of the effort to incentivize the cybersecurity industry has come about because of an executive order signed by President Barack Obama Feb. 12 that directs federal agencies to develop voluntary cybersecurity standards for private-sector industries and propose new mandates if needed. It was aimed at helping state and local governments protect critical infrastructure controlled by Web-based technology.

Widely publicized data breaches over the past couple of years fueled the government's effort. They include:

- In March 2012, NASA shut down a large database and sent warnings to its employees after a laptop stolen from a car was hacked, revealing personnel data as well as technology

- In 2012, hackers got into the U.S. Navy system that tracks personnel moves and compromised private data on 200,000 sailors and their family members.
- In 2012 a Washington state website was hacked, revealing hundreds of thousands of Social Security and driver's license numbers of state residents.

Tax Breaks

The trade group Council for Community and Economic Research keeps track of tax incentives across the 50 states and offers businesses incentive comparisons. The group's website notes that there are more than 1,600 incentive programs across the country sponsored by different states and localities. and the cybersecurity credit is just one focus.

New York and New Jersey, for example, are competing for business—this time in the financial sector, tax expert Colavito said. "States are being aggressive on both sides," he said.

He also pointed out that despite the tax break for investing in Maryland, the state still stands to gain if a firm is successful, because it will then pay taxes to the state. The break is a "refundable credit," however, so it applies even if the company never makes money.

O'Malley and Michigan Gov. Rick Snyder headed up the National Governors Association's Resource Center on Cybersecurity, and promoted the topic at the NGA's winter meeting in February.

Whether the return on investment for the state is profitable is an open question. According to a study in April 2012 by The Pew Charitable Trusts, states have a spotty track record on following up on the results of their economic development tax incentives. While every state has some kind of tax incentive, and many of the states have several, not all are successful at determining their outcomes.

Pew, *Stateline's* parent organization, rated the states and the District of Columbia on how well they are measuring the economic benefits and costs of their tax incentives. Thirteen states were rated as "leading the way," 12 got "mixed results" and 26 were "trailing behind."

49. How the National Guard Is Protecting Cybersecurity*

COLIN WOOD

A $46 billion annual business of protecting infrastructure from cyberattacks largely revolves around the federal government. But within the past year, efforts have ramped up to bring federal-level cybertools and resources to state and local governments—and the National Guard may be the vehicle for driving that collaboration.

The feds have been trying to go at cybersecurity alone for years, but they're finally coming around and including states and localities, said Heather Hogsett, director of the National Governors Association's (NGA) homeland security and public safety committee. Last year, the NGA backed a bill called the Cyber Warrior Act of 2013, which would have directed the Department of Defense to establish "Cyber and Computer Network Incident Response" teams composed of National Guard members in each state.

Although the measure failed to pass last year, it drew attention to the issue. And state-level efforts—like the National Guard's cyberteam in Washington state—continue to expand the Guard's cyberprotection role.

Congress is hearing from lower governments on the cyberissue. Last September, Michigan Gov. Rick Snyder briefed Congress on the NGA's cybersecurity efforts, emphasizing the importance of state government's growing role. During the event, Snyder released a paper called Act and Adjust: A Call to Action for Governors for Cybersecurity, a-six page document outlining recommendations for states that want to improve their cybersecurity. Snyder also released a piece of software, now being tested in Michigan and Maryland,

---

*Originally published as Colin Wood, "How the National Guard Is Protecting Cybersecurity," *Government Technology*, March 3, 2014. Reprinted with permission of the publisher.

that allows governors to see an overview of their state's cybersecurity environment.

"Governors are very focused on cybersecurity, and we at NGA are trying to provide them with any tools and resources available to help them better protect critical fiber infrastructure and assets that reside in their state," Hogsett said. Bringing the nation's governors into the world of cybersecurity would be mutually beneficial for states and the federal government, and it makes sense for the guard to fill that role, she said.

"The National Guard is unique in the fact that it can serve both the governors and the president. It's the only military service that can do that," she said. "Both the federal government and states have pretty widely put out there that there's a shortage of trained, qualified personnel to help perform cybersecurity functions." And the National Guard is in a perfect position to recruit skilled private-sector professionals to assist the government with cybersecurity. Concerned IT professionals wouldn't need to join the guard, Hogsett said—they could just help during their free time because the National Guard has the ability to do that.

The National Guard is trusted, well known and cost-efficient, she added. "For the cost of a single active-duty soldier, you can essentially provide two to three National Guard members," she said. "It's a really solid resource that we believe can and should be better leveraged."

The timeline on this isn't five or 10 years, she said—this is more likely something that could happen in the next 12 to 18 months.

South Carolina learned its cybersecurity lesson the hard way in 2012. The state's Department of Revenue was the target of an attack that exposed millions of Social Security numbers, thousands of credit card numbers, along with lots of other personal information. The months-long ordeal cost South Carolina at least $14 million and damaged the government's reputation with citizens, making the state just one victim in a string of large attacks to hit the public sector over the past few years.

At the very least, states need to have a cybersecurity emergency preparedness plan, recently retired South Carolina CIO Jimmy Earley said. "You do not want to go through the process of thinking through what needs to happen and who needs to do what, while you're reacting to it," he said. "You need to have that plan and that process nailed down before you actually have to react to something like this."

South Carolina contracted with Deloitte to help resolve its security issues last March, Earley said. They've assessed three agencies, will assess 15 more agencies and are establishing a security framework and governance model for the whole organization.

"As a state, we have a very decentralized model for using IT," Earley said. "We have 70-plus agencies in the state, and most agencies procure, manage and implement IT independent of each other and really outside of any central framework or structure. Each agency is doing the best they can, making decisions about security controls that need to be in place, and how to best manage security for their agency. That environment is ripe for problems. What we really felt we needed was a simpler approach to manage security in the state."

Working together and sharing information is one of the best things organizations can do in the face of cyberthreats, Earley said.

South Carolina isn't unique, said Doug Robinson, executive director of NASCIO. "From the CIO perspective, there is a definite gap in terms of a documented response and recovery plan," he said, and many organizations are still figuring out what their roles are supposed to be in the world of cybersecurity. Clearly defined roles is one of the things the NGA is trying to establish as governments at all levels determine what their jobs are in the national effort to protect computer networks.

Roles in cybersecurity are changing and many of the changes are for the better, Robinson said. State CIOs have in recent years been allowed security clearance in order to access more information held by federal agencies like the Department of Homeland Security (DHS), but the National Guard could help further bridge the gap between local and federal government, giving states and localities more autonomy and knocking down some of the institutional barriers.

In states like Washington, the guard has a head start on demonstrating its ability to coordinate cybersecurity activities and response. The National Guard adjutant general, a position currently held by Bret Daugherty, also serves as state homeland security adviser and director of emergency management, three roles that allow one individual to bridge jurisdictions and simplify command of federal resources and the Washington State Fusion Center, while leading the state's cybersecurity team, said Kelly Hughes, director of plans and programs at the Washington Air National Guard.

"If a utility gets hacked really badly, they reach out to the Department of Homeland Security, they can get teams or support to help them mitigate it and figure out what happened," Hughes said. "Before, they would just go direct to those agencies by themselves. Now, they go through the state military department, so we coordinate those efforts."

Coordinating the state's efforts through a central authority has the advantage of increased awareness and shared resources, Hughes said. It also gives them the opportunity to work with the FBI and the state fusion center so they can reach out to other organizations that may have been affected by an attack but didn't know it.

For the last few years, the Washington National Guard has been running cyberexercises with technical help from the DHS, Hughes said, but last fall the state was scheduled to test its cyberincident response plan without input from the federal agency. "We're going to test that plan with a group of policy folks from state, local and hopefully some local private industry as well to say, 'If we did have [an incident], how would we respond? Bring your Rolodex. How many smart guys can we call off our own phones before we have to ask somebody else to come in and help us?'"

The National Guard is a great partner, said Washington state CIO Michael Cockrill. "Security is my No. 1 focus overall," he said. "Generally when someone asks me what my top three foci are, I say security, security and security. And then we talk about No. 4 and 5. … The security landscape on a global basis is changing so fast that it takes a constant effort to keep up with it, and it has to be the highest priority of the state to keep citizens' data safe."

Using the National Guard for testing cybersecurity is great not just because it has access to federal resources and offers a more centralized command structure, Cockrill said, but it's a cost savings to the state too. Using an outside organization for such testing would be costly and less secure. "We can keep it all in-house, and it's going to be much more streamlined in doing this super-critical penetration testing."

In Michigan, the National Guard applied for funding to begin an interstate network of cyber-range facilities that would allow for public and private industry to participate in joint exercises without needing security clearance. Existing federal projects, like the Defense Advanced Research Projects Agency's (DARPA) $110 million National Cyber Range, are helpful, said Brig. Gen. Michael Stone of the Michigan National Guard, but only to those with top security clearance. An interstate network of cyber-range facilities would provide valuable research and analysis of cybervulnerabilities for state and local operators of critical infrastructure.

"There are folks who work at the federal level, policy makers, who believe the domain of cyber falls entirely on the federal government," Stone said. "The problem is that requires perfect resources and perfect execution by the federal government. And how perfect is federal government execution all the time?"

It doesn't make sense to put federal agencies in charge of critical infrastructure such as power grids and dams, Stone said, because that's not who's operating them. "Eighty percent of all critical infrastructure is privately owned," he said. "And 85 percent of all people operating networks for critical infrastructure are civilians, nonfederal government."

Not everyone favors more local control, though. Gartner Analyst Lawrence Pingree said the fed-centric model has some strengths. "I am unconvinced that the state and muni level is the right approach since the amount of spend should be more centralized and administered in a similar fashion to support efficient deployment of capital," he said. "Also, one major problem government has is that it is often unwilling to pay the appropriate salary levels that security practitioners can demand in the private sector, significantly limiting their ability to execute or retain talent once it is developed."

But Stone contends that a network of state cyber-ranges would be both valuable and economical. He said the cost for establishing each facility in the network is in the hundreds of thousands of dollars, as opposed to the millions spent by the federal government. "The dollar figure to stand up hubs is really the cost of running fiber optic to the buildings we want, which is about $50,000 a mile," he said. "Once you're there, it's really the human capital cost."

The Michigan National Guard partnered on the initiative with a handful of other organizations, including the California National Guard; California Polytechnic State University, San Luis Obispo (Cal Poly); and Michigan's Merit Network, a high-performance network linking universities, K–12 schools, government agencies and nonprofits in the state. Electricore, a nonprofit group of public and private organizations established by DARPA to develop advanced technology, applied for a U.S. Department of Energy grant on behalf of the team's members.

Michigan opened a public-private cyber-range in 2012. Other participants will include Cal Poly and major universities in Michigan. Some of the first hubs will be military bases and academies in Michigan. Stone said he is also in talks with the National Guard Bureau in Little Rock, Ark., and organizations in Kansas. The idea, he said, is to cast a wide net while also creating a culture of cybersecurity awareness. "We're going to need special guardsmen with civilian skill sets. We're going to need recent college graduates; we're going to need an abundance of IT experts to really be able to surge, to overcome those problems."

Michigan CIO David Behen views the cyber-range initiative as a way to strengthen his state's cyber-readiness and spur economic development.

"I believe that the Michigan cyber-range, through a public-private partnership, is the exact model we need to build a cybersecurity industry here in Michigan," he said. "That's what we're really excited about. How can we draw entrepreneurs? How can we use cybersecurity in a positive way around economic development?"

50. Cybersecurity and Local Governments in the United States*

WILLIAM HATCHER

A few years ago, I was part of a team that helped local governments in eastern Kentucky offer e-government services by having functional websites, online bill payment and other internet based technologies to serve citizens. From our team's conversations with local officials, two issues were identified as barriers: (1) the cost of e-government services and (2) security concerns.

The cybersecurity events of this year have most likely increased the privacy concerns of the public servants in Eastern Kentucky and throughout the nation. Local governments should not allow the barriers of cost and security to keep them from improving the effectiveness of their administrative actions by making use of modern technologies. The problem is local governments, especially small cities and counties, lack the administrative capacity and expertise to properly address these barriers. Thus, public administration needs to be active in helping communities find financing for e-government services and ensuring those services are secure.

At Augusta University, our MPA program serves a community labeled as an emerging cybersecurity and information technology hub. The community's Fort Gordon is the U.S. Army's location for its Cyber Command. With this destination, the Augusta Metro is experiencing a growth of cyber-related industries and jobs. The community's role in cyber is gaining attention. *Fortune* included the metro on its list of "7 cities that could become the world's

*Originally published as William Hatcher, "Cybersecurity and Local Governments in the United States," *PA Times*, May 9, 2017. Reprinted with permission of the publisher and author.

cybersecurity capital." And *Forbes* has included commentary on how the city is "becoming a model for tech innovation." Besides Augusta, the other global cities are: Atlanta, Washington, D.C., Silicon Valley, London, Tel Aviv, and Boston.

To serve our community, the Augusta University's MPA is focused on helping local nonprofits and public agencies prepare and take advantage of the potential for Augusta to be the cybersecurity capital of the world. We're learning that MPA and other public affairs programs should play a role in helping local governments provide effective and secure e-government services. Programs can do so by focusing on:

- Securing funding through grant writing assistance;
- Conducting applied research for local partners on the efficacy of e-government services; and,
- Providing training on how to provide effective and secure e-government services.

Funding

Public administration programs can offer grant writing assistance to local governments in order to help them obtain funding for e-government services. Programs can work with state governments and federal agencies to help identify funding. MPA faculty, students and staff can also help local governments evaluate the website and e-government services provided by private contractors.

Applied Research

Providing active applied research is perhaps the most important assistance public affairs programs can offer to strength the e-government capacity of local governments. The applied research can include a variety of topics and projects, such as showing the efficacy of online bill payment, the usefulness of social media sites and the value of promoting community assets on the web.

Augusta University's MPA program is a partner in applied research projects that are helping build e-government capacity in our community. We're partnering with the Augusta University Cyber Institute to identify the supply and demand cyber-related occupations in the Augusta Metro. The study includes analysis of labor statistics and a survey asking local employers (businesses, nonprofits and public agencies) to identify their cyber-related employment needs for the future and the type of education future employees will need to be competitive for the positions. Furthermore, we're learning the cyber needs of local agencies, so we can serve them better in our community outreach.

Training

Public affairs programs can help build administrative capacity for e-government by offering robust training programs. At Augusta University, our program, due to funding from our Cyber Institute, is offering free online e-government trainings for public agencies. The training includes three modules:

- An overview and definition of e-government services
- How local governments use e-governments services
- How e-government services can be kept secure

We're hoping this training helps public agencies provide effective and secure e-government services.

With Facebook having close to 2 billion active users and many people getting most of their news via social media, public administration needs to do a better job helping communities offer e-government services and providing e-governance. Since the advent of the internet, private firms have been more innovative in providing online services. However, local governments are making progress, especially large cities and manager-council forms of government. Our field needs to be more active in helping medium sized and rural communities provide services through internet based technologies. We can do this by helping communities secure funding, conducting applied research and providing training.

51. The Top 18 Security Predictions for 2018*

DANIEL J. LOHRMANN

Abraham Lincoln once said, "The best thing about the future is that it comes one day at a time."

Winston Churchill once said, "If you're going through hell, keep going." And, "Never, never, never give up."

As we look back at top cyber stories and security trends in 2017, these wise words from fearless leaders who have gone before us certainly apply to cybersecurity and the new 21st-century challenges confronting our world in 2018.

What's HOT and Likely Getting HOTTER in 2018?

Last year we started with, "You ain't seen nothing yet!"

Hold on! 2018 will be even worse online, if these global security experts are correct.

No doubt, more sophisticated hacker tricks, phishing attempts and data breaches are coming.

What are the most common security predictions for next year? New forms of malware, more expensive ransoms as more ransomware hits more organizations, Internet of Things (IoT) device problems at home, AI and machine learning gone astray (as a cyberweapon), cryptocurrency problems, cloud computing breaches and plenty more of everything we already saw in 2017.

---

*Originally published as Daniel J. Lohrmann, "The Top 18 Security Predictions for 2018," *Government Technology*, January 4, 2018. Reprinted with permission of the publisher.

Almost everyone is talking about the huge impact of GDPR in 2018—some think the fines will wait for later after lawsuits will be filed, but most see a major shake-up coming for companies' policies and procedures as a result of the new European privacy rules.

Other common cyberpredictions include increased scope and impact from DDOS attacks, the number of cybercriminals (and crimes) increasing, continued shortages of qualified security professionals—with new attempts to deal with the staffing problems, popular (and easy to use) home devices (such as Amazon Echo) getting hacked in new ways and much more nation-state hacking.

In addition, the election hacks, hacktivism and business email compromised (CEO fraud) show up on many lists as likely items that will expand in the coming year.

Why Take the Time to Understand Cybersecurity Industry Predictions?

There's no doubt that security predictions are exploding and cover a very wide range of technology, physical security and Internet of Things (IoT) topics around the world. The breadth and depth of industry involvement in this cyber forecasting process even exceeds previous years, which is truly remarkable and shows the dramatic growth of the security industry as a whole.

So why take the time to go through these lists? I addressed this topic in detail back in 2016 for CSO Magazine in this piece: Why more security predictions and how can you benefit? I started by saying that Americans love baseball, hot dogs, apple pie and predictions. I also predicted that more security predictions would be coming—and I certainly nailed that cybersecurity trend.

But beyond just a fun end-of-the-year exercise, there is immense value for individuals and companies as they plan their future strategies. Here's an excerpt of a few of the benefits to understanding what experts think may be coming soon:

- Gain industry knowledge, understand overall trends and expand your horizons beyond one stovepipe or topic. Security predictions help you understand industry trends and help you grow in your knowledge—if you do your homework and read the supporting research that usually comes from major vendors. Remember that the actual date the event happens is less important than trends, patterns and even repetition of an item. …

- Use the free advice, direction, insights and annual reports provided by many to respond to the expected cyberthreats.
- Use predictions as an opportunity to educate others. Get the word out on cybersecurity—whether that is to your company, your family or your community group. Are you bringing problems or solutions? We claim we want to educate end users on cybersecurity, so educate!

Quick Reminders

No doubt, there are some leftover (very similar) predictions from the past few years. There is also the annual chorus of: "Will this be the year for a Cyber Pearl Harbor or a Cyber 9/11 that brings down critical infrastructure for a section of the country?"

To get a full sense of the breadth and depth of security industry prediction lists and forecasts, I recommend going back in time and reviewing some of the previous security prediction roundups from 2015, 2016 and 2017 to help keep score on prognosticators. Our analysis process has not changed much in the many years since we started, and all decisions are made independent of company or magazine influence.

For more details, I encourage you to go to the prediction details by clicking on the hyperlinked report and/or visit the specific website and download the full white papers to get more details on these security trends and 2018 predictions lists. Many of these predictions have longer explanations as to why this will happen (with more data to share.) Be aware that some vendors may require you to register (often for free) to get their full prediction report.

So now we're ready to move on to the best (most complete) security prediction list for 2018, ranked from 1–18 using my vendor-agnostic rating system, along with honorable mention and late-arriving prediction lists.

Detailed Prediction Reports by Source

1. **Trend Micro** takes the top prize for again having an impressive, well rounded set of predictions. The Trend Micro theme is "Are You Ready for Paradigm Shifts," and here are their top predictions:

- In 2018, digital extortion will be at the core of most cybercriminals' business model and will propel them into other schemes that will get their hands on potentially hefty payouts.
- The ransomware business model will still be a cybercrime mainstay

in 2018, while other forms of digital extortion will gain more ground.
- Cybercriminals will explore new ways to abuse IoT devices for their own gain.
- Global losses from Business Email Compromise scams will exceed US$9 billion in 2018.
- Cyberpropaganda campaigns will be refined using tried-and-tested techniques from past spam campaigns.
- Threat actors will ride on machine learning and blockchain technologies to expand their evasion techniques.
- Many companies will take definitive actions on the General Data Protection Regulation (GDPR) only when the first high-profile lawsuit is filed.
- Enterprise applications and platforms will be at risk of manipulation and vulnerabilities.

2. **Symantec** had another outstanding set of predictions for 2018 on a wide range of topics:

- Blockchain Will Find Uses Outside of Cryptocurrencies but Cyber criminals Will Focus on Coins and Exchanges
- Cyber Criminals Will Use Artificial Intelligence (AI) & Machine Learning (ML) to Conduct Attacks
- Supply Chain Attacks Will Become Mainstream
- File-less and File-light Malware Will Explode
- Organizations Will Still Struggle With Security-as-a-Service (SaaS) Security
- Organizations Will Still Struggle With Infrastructure-as-a-Service (IaaS) Security—More Breaches Due to Error, Compromise & Design
- Financial Trojans Will Still Account for More Losses Than Ransomware
- Expensive Home Devices Will Be Held to Ransom
- IoT Devices Will Be Hijacked and Used in DDoS Attacks
- IoT Devices Will Provide Persistent Access to Home Networks

3. **Watchguard Technologies** —I really like Watchguard's presentation of predictions again! In fact, I would say that their online videos and infographics may be my favorite this year. However, their actual predictions seemed rather mainstream and offered no huge surprises. Very solid list though:

- Cryptocurrency Crash
- Wi-Fi Hacking

- Increased Adoption of Corporate Cyber Insurance
- IoT Botnets Force New Regulations
- Linux Attacks Will Double
- Multi-factor Authentication
- Hack Election Machines

4. McAfee—McAfee forecasts developments in adversarial machine learning, ransomware, serverless apps, connected home privacy, and privacy of child-generated content. Here are some details:

- McAfee Labs predicts an adversarial machine learning "arms race" between attackers and defenders
- Ransomware to evolve from traditional PC extortion to IoT, high net-worth users, and corporate disruption
- Serverless Apps to create attack opportunities targeting privileges, app dependencies, and data transfers
- Connected home devices to surrender consumer privacy to corporate marketers
- Consumer apps collection of children's content to pose long-term reputation risk

5. FireEye offers excellent predictions, but requires you sign up for the full report (which is free). Nevertheless, this interview with FireEye executive leadership, including their CEO Kevin Mandia, is eye-opening regarding 2018 predictions:

In the Indo-Pacific region, FireEye said, China and neighboring countries are still continuing political disputes, especially with India, South Korea, Japan, the Philippines, Vietnam and other South-east Asian countries.

"Therefore, unorganized 'hacktivism' attacks as a response to these political tensions within and against these countries is expected to continue and possibly rise throughout the new year," the company warned.

According to FireEye, it observed an increase in non–Chinese and non–Russian APT groups in 2017 and expects to discover more in 2018. Ransomware is expected to rise in 2018, especially as administrators are slow to patch and update their systems.

Other popular techniques that will continue to be used in 2018 are strategic web compromises and spear phishing, especially in targeted attacks. We also expect to see many more destructive worms and wipers, the cyber security firm noted.

6. Kaspersky —Offers detailed cyberthreat forecasts in each major sector. For example, their financial predictions include:

- Cryptocurrency—in vogue in the cybercriminal world
- Speed increases danger

- Fraud as a service
- Other Kaspersky predictions about auto, connected health, industrial security and cryptocurrencies can be found at this excellent SlideShare.

7. Palo Alto Networks—Human safety and security will be added to confidentiality, integrity and availability, according to Palo Alto Networks.

8. Forcepoint—Offers eight different areas of concern for the year ahead and five predictions for 2018.

- An increasing amount of malware will become MitM [Man in the Middle]-aware.
- IoT is not held to ransom but instead becomes a target for mass disruption.
- Attackers will target vulnerabilities in systems which implement blockchain technology.
- A data aggregator will be successfully breached in 2018 using multiple attack methods.

9. Imperva—Offers Their Top 5 Trends That IT Pros Need to Think About:

- Massive Cloud Data Breach
- Cryptocurrency Mining
- Malicious Use of AI/Deception of AI Systems
- Cyber Extortion Targets Business Disruption
- Breach by Insiders

10. Forrester—As always, Forrester offers some unique and thought-provoking predictions for 2018:

- Governments will no longer be the sole providers of reliable, verified identities
- More IoT attacks will be motivated by financial gain than chaos
- Cybercriminals will use ransomware to shut down point of sale systems
- Cybercriminals will attempt to undermine the integrity of US 2018 midterm elections
- Blockchain will overtake AI in VC funding and security vendor road maps
- Firms too aggressively hunting insider threats will face lawsuits and GDPR fines

11. Webroot—Excellent, wide assortment of predictions on topics ranging from ransomware to breaches to biometric security to government security to the infosec job market.

- Backups will not prove enough to stop ransomware as hackers find ways to subvert this strategy.
- Consumer fightback—2018 will see major a major backlash (maybe class action lawsuits) from consumers, requiring more regulations around data protection especially in the U.S.
- An increase in nation state cybersecurity breach activity as "cold war" like activity continues to escalate. Where countries and organizations (e.g., ISIS) will actually invest more into both defensive and offensive tech and skills to gain access to information that can be leveraged in numerous ways. I think we have only seen the early days of what's possible and likely here.
- Discoveries of election meddling and social media tweaking will be an economic drag on some of the biggest tech giants in the industry—and be cause for further scrutiny on securing devices, networks, and communications channels and verifying identity. The tradeoffs between free speech and open digital access and convenience will become ever more apparent.
- State-sponsored service breach of critical infrastructure leading to loss of life and an extended timeframe to return to normal operations

12. Gartner—Gartner again offers 10 strategic predictions (via PC Magazine) that cover the next few years (through 2022). Here are a few of the security-related predictions from Gartner:

- By year end 2020, the bank industry will derive 1 billion dollars of business value from the use of blockchain-based cryptocurrencies.
- Through 2022, half of all security budgets for IoT will go to fault remediation, recalls, and safety failures, rather than to protection. Most organizations don't have a budget for IoT security now, but they will need to add one, [Gartner Fellow Daryl] Plummer said. By 2019, IoT security incidents will make the nightly news.
- Through 2021, AI-driven creation of "counterfeit reality," or fake content, will outpace AI's ability to detect it, fomenting digital distrust.

In early December 2017, Gartner issued a forecast that worldwide enterprise security spending will rise 8 percent in 2018 to $96.3 billion.

13. Sophos—Offers details on malware likely coming in 2018.

And their PDF offers excellent details and a new malware forecast. They write: "In this report, we review malicious activity Sophos Labs analyzed and protected customers against in 2017 and use the findings to predict what might happen in 2018.

The malware we protect customers from transcends operating systems. Ransomware in particular targets Android, Mac, Windows and Linux users alike. (Android phones run a modified version of Linux.) Four trends stood out in 2017 and will likely dominate in 2018."

- A ransomware surge fueled by RaaS [ransomware as a service] and amplified by the resurgence of worms;
- An explosion of Android malware on Google Play and elsewhere;
- Continued efforts to infect Mac computers; and
- Ongoing Windows threats, fueled by do-it-yourself exploit kits that make it easy to target Microsoft Office vulnerabilities

14. Zscaler —Ten interesting predictions, including this unique and creative one:

"We will see targeted attacks on digital assistants."

It seems that every major tech company is now convinced that digital assistants (Alexa, Siri, Cortana) embodied as smart speakers (Amazon Echo, Apple HomePod) are the future of human-computer interaction. These devices are now mainstream and have become much more than just a convenient way to learn about today's weather or get the latest sports scores.

15. IBM—Offers interesting predictions, with the first two items being somewhat different than many other lists:

- AI Versus AI
- Africa Emerges as a New Area for Threat Actors and Targets
- Identity Crisis
- Ransomware Locks Up IoT Devices
- Finally Getting Response Right

16. eWeek says that "Cars Steal Innovation Spotlight From Smartphones"

- Autonomous vehicles: "In the world of autonomous vehicles, we predict we are going to see much more incremental progress, and a slow and steady shift toward collaboration. Right now, it seems many are quick to imagine that a utopia of fully autonomous vehicles is just around the corner; however, the reality is that right now our algorithms just understand how humans drive with humans. Given this, our algorithms will need to evolve to better understand the nuances of how humans drive with semi- and fully autonomous vehicles; how various models from different manufacturers interact with each other on the road; and in diverse environments, infrastructure and weather conditions."

- Cutting the car: "Just as cable television users are cutting the cord in favor of streaming, this rise of shared mobility will lead some consumers to cutting the car. Personal car ownership will decrease over the years as alternative types of auto mobility flourish, and we project that Europe specifically will reach peak car by 2020. How soon we will see these shifts occur elsewhere remain to be seen, but it's safe to say that personal mobility will look drastically different a decade from today."

Update: eWeek also released this helpful slide show of 18 cyber security trends that organizations should be aware of heading into 2018. They follow my "18 for 2018" model in this annual cybersecurity prediction blog. (Imitation is the greatest form of flattery, so thanks.)

17) Checkpoint sticks to a few unique items in their forecast:

- Legitimate Organizations Caught Hacking
- Will Cryptocurrencies Be Regulated?
- Governments Deploying Cyber-Armies to Defend Their Citizens and Borders

18. White Hat Security—Last year, Ryan O'Leary said, "Nothing will change. Companies will continue to get breached because of simple vulnerabilities." Unfortunately, my prediction was correct, but that's no surprise." This is still a good prediction for 2018.

New this year: "...More and more companies will start adopting the DevSecOps process and bring the Development, Security and Operations teams together. We've seen this work with companies and we know it reduces both the number of vulnerabilities introduced, and also the time to fix those vulnerabilities. By making one team with the mission of fast, secure, and stable code we ensure that these teams no longer have competing priorities which hinder secure releases...."

Final Thoughts

I did not see very much missing this year on these prediction and forecast reports, but the Winter Olympics in S. Korea and FIFA World Cup (soccer) in Russia are noticeably absent. Of course, we also have the Super Bowl, World Series, March Madness and other major sporting events that could be disrupted.

There were plenty of people predicting critical infrastructure disruptions, but no one really sticking their necks out to say a major critical system failure (such as a dramatic regionwide or nationwide power outage or the significant loss of life because of hospital systems failure) is likely due to hacking.

Still, I agree with Bruce Schneier that regulation is coming for IoT when someone clearly dies from a cyberattack. Will 2018 be the year? Perhaps.

In conclusion, here's one more quote from Abraham Lincoln that still applies as we head into 2018:

"The best way to predict your future is to create it."

52. Artificial Intelligence Cyber Attacks Are Coming—But What Does That Mean?*

JEREMY STRAUB

The next major cyberattack could involve artificial intelligence systems. It could even happen soon: At a recent cybersecurity conference, 62 industry professionals, out of the 100 questioned, said they thought the first AI-enhanced cyberattack could come in the next 12 months.

This doesn't mean robots will be marching down Main Street. Rather, artificial intelligence will make existing cyberattack efforts—things like identity theft, denial-of-service attacks and password cracking—more powerful and more efficient. This is dangerous enough—this type of hacking can steal money, cause emotional harm and even injure or kill people. Larger attacks can cut power to hundreds of thousands of people, shut down hospitals and even affect national security.

As a scholar who has studied AI decision-making, I can tell you that interpreting human actions is still difficult for AI's and that humans don't really trust AI systems to make major decisions. So, unlike in the movies, the capabilities AI could bring to cyberattacks—and cyberdefense—are not likely to immediately involve computers choosing targets and attacking them on their own. People will still have to create attack AI systems, and launch them at particular targets. But nevertheless, adding AI to today's cybercrime and cybersecurity world will escalate what is already a rapidly changing arms race between attackers and defenders.

---

*Originally published as Jeremy Straub, "Artificial Intelligence Cyber Attacks Are Coming—But What Does That Mean?," *The Conversation*, August 27, 2017. Reprinted with permission of the publisher.

Faster Attacks

Beyond computers' lack of need for food and sleep—needs that limit human hackers' efforts, even when they work in teams—automation can make complex attacks much faster and more effective.

To date, the effects of automation have been limited. Very rudimentary AI-like capabilities have for decades given virus programs the ability to self-replicate, spreading from computer to computer without specific human instructions. In addition, programmers have used their skills to automate different elements of hacking efforts. Distributed attacks, for example, involve triggering a remote program on several computers or devices to overwhelm servers. The attack that shut down large sections of the internet in October 2016 used this type of approach. In some cases, common attacks are made available as a script that allows an unsophisticated user to choose a target and launch an attack against it.

AI, however, could help human cybercriminals customize attacks. Spearphishing attacks, for instance, require attackers to have personal information about prospective targets, details like where they bank or what medical insurance company they use. AI systems can help gather, organize and process large databases to connect identifying information, making this type of attack easier and faster to carry out. That reduced workload may drive thieves to launch lots of smaller attacks that go unnoticed for a long period of time—if detected at all—due to their more limited impact.

AI systems could even be used to pull information together from multiple sources to identify people who would be particularly vulnerable to attack. Someone who is hospitalized or in a nursing home, for example, might not notice money missing out of their account until long after the thief has gotten away.

Improved Adaptation

AI-enabled attackers will also be much faster to react when they encounter resistance, or when cybersecurity experts fix weaknesses that had previously allowed entry by unauthorized users. The AI may be able to exploit another vulnerability, or start scanning for new ways into the system—without waiting for human instructions.

This could mean that human responders and defenders find themselves unable to keep up with the speed of incoming attacks. It may result in a programming and technological arms race, with defenders developing AI assistants to identify and protect against attacks—or perhaps even AI's with retaliatory attack capabilities.

Avoiding the Dangers

Operating autonomously could lead AI systems to attack a system it shouldn't, or cause unexpected damage. For example, software started by an attacker intending only to steal money might decide to target a hospital computer in a way that causes human injury or death. The potential for unmanned aerial vehicles to operate autonomously has raised similar questions of the need for humans to make the decisions about targets.

The consequences and implications are significant, but most people won't notice a big change when the first AI attack is unleashed. For most of those affected, the outcome will be the same as human-triggered attacks. But as we continue to fill our homes, factories, offices and roads with internet-connected robotic systems, the potential effects of an attack by artificial intelligence only grows.

53. FCC Group on 5G Deployment Short on Local Input*

MARY ANN BARTON

Some members of an FCC-appointed committee looking at how to speed deployment of 5G telecommunications services say the makeup of their group is "going to go nowhere" without more input from local government.

The FCC's Broadband Deployment Advisory Committee was formed to make recommendations, due in November, on how to streamline local government rules for siting small cell nodes and towers to accelerate the rollout of next-generation telecom services.

While there are no standards yet for 5G mobile networks, members of the telecom industry say to expect speeds 10 times faster than current 4G networks.

The committee, which has met publically two times, is mainly composed of telecom industry executives and members of conservative think tanks, according to the Center for Public Integrity. More than 60 local government officials applied to be on the 30-member committee, only two were initially chosen by the FCC, with two more added later.

The committee's scant local government representation has drawn criticism from within its ranks.

"This is a serious risk right now—we have a lot of groups who are concerned that they're not at the table, and if they don't feel included ... it's those groups we want to adopt model codes," said committee member David Don, vice president of regulatory policy for Comcast, at the committee's second

---

*Originally published as Mary Ann Barton, "FCC Group on 5G Deployment Short on Local Input," *NACo County News*, Sep. 18, 2017. Reprinted with permission of the publisher.

public meeting. "I think, just at the end, if we present a model code ... and they feel they've had an insufficient amount of input, it's going to go nowhere.... One way or another, we're going to need their buy-in."

"If [local governments] don't buy in, this is going to fail and we've all wasted our time," University of Pennsylvania Law School professor and committee member, Christopher Yoo said at the meeting.

The FCC committee has identified 189 local government "barriers" to deployment of broadband but an FCC spokesman said the information could not be shared until the group's report is made public in November.

Shireen Santosham, chief innovation officer for San Jose, Calif., told the group that local governments have several concerns, including caps on fees that are redeployed to help bring telecom services to areas that are underserved.

Montgomery County, Md., Councilmember Hans Riemer said he applied to be on the BDAC committee and although he was not accepted, he was invited to a closed-door meeting Aug. 29 of the committee's Model Code for Municipalities Working Group; the FCC said the group is working to "develop fair and reasonable guidelines for the use of public assets to ensure the best overall outcome for all residents." It is charged with drafting a model code for municipalities to accelerate broadband deployment to be voted on by all members of the committee at a public meeting in October or November.

"I really tried to convey to them that we need a collaborative approach to siting," Riemer said. "If we get steamrolled, the whole effort is going to get bogged down."

"There is no one who agrees more than local government that we want to help deliver better services to our residents," he said. "But getting government out of the way is not part of the solution."

Paralleling the group's work, the FCC announced two rulemaking procedures in the spring that look into the 5G rollout issue; but the agency has told the committee that its recommendations may or may not have any bearing on the rulemakings, which "aim to do things like speed decisions by state or local governments on tower siting applications and ease access to utility poles," FCC Commissioner Michael O'Rielly told an audience in Virginia last month.

While NACo supports next-generation telecommunications, it opposes preemption of local authority as well as the preemption of the county role in wireless communications facilities siting.

Members of the public may submit comments to the BDAC in the FCC's Electronic Comment Filing System at www.fcc.gov/ecfs. Comments to the BDAC should be filed in GN Docket No. 17–83.

Meanwhile in State Capitals, Local Zoning and Licensing Regulations Are Under Siege

Arizona: HB 2365 was passed in April and became law Aug. 9. It requires counties to establish rates, fees and terms for the installation, modification or replacement by a wireless provider of a utility pole located in a right of way; the co-location of a small wireless facility in a right of way and the co-location of a small wireless facility on a county utility pole.

The Maricopa County Department of Transportation must now allow "small wireless facilities" to be installed on its traffic-signal poles, MCDOT spokesperson Nicole Moon recently told the *North Phoenix News.* The units are not to exceed roughly the size of an end table. Small equipment boxes will typically be mounted on the poles or nearby.

California: Counties are urging Gov. Jerry Brown not to sign SB 649, which passed Sept. 14. The California State Association of Counties estimated a loss of $100 million a year due to decreased fees. CSAC said "...locals wouldn't be able to require in-kind public services (such as Wi-Fi access or connecting civic amenities to fiber) in exchange for the use of these publicly-owned structures. Evidence that the bill is even needed is lacking, as jurisdictions around the state have already worked collaboratively with providers to install similar antennas on publicly owned poles."

In a column published in *The Sacramento Bee*, CSAC called the then-proposed legislation "a triple rip-off: Neighborhoods lose flexibility to address visual blight, taxpayers get shorted by limits on lease revenues and residents still pay the same high prices for their wireless services. At the same time, these providers still won't serve many of California's disadvantaged communities."

Delaware: Gov. John Carney (D), signed HB 189 last month; The new law enables carriers and their partners to apply to place small cells on public rights of way directly through the state's department of transportation.

Florida: New law HB 687 signed June 23 by Gov. Rick Scott (R) caps attachment fees at $150; Florida counties stand to lose as much as $145 million in fees annually, according to some estimates. Six counties had issued moratoriums on the placement of the equipment. "It really takes away our ability to regulate cellphone facilities on our right of way," Seminole County, Fla. Commission Chairman John Horan recently told the Orlando Sentinel. "These are in the right of way. They belong to the public, and these are private cellphone companies using the right of way." Counties have until Jan.1, 2018 or three months from their first application to create local ordinances that adhere to the new law.

New York: SB 6687 directs the public service commission to prohibit attachment of wireless equipment to existing utility poles in certain circumstances.

Virginia: Gov. Terry McAuliffe (D) signed SB 1282 allowing telecom companies to attach "small cell facilities" to towers, buildings, utility poles, light poles, flag poles, signs and water towers. Under the new law, companies pay $100 for up to five small cell facilities on a permit application. The new law also prohibits local governments from adopting a moratorium on considering zoning applications.

New small cell wireless legislation was also signed into law in Minnesota, North Carolina, Ohio and Texas where cities have telecom siting authority.

54. Thinking Strategically in the Cyber Domain*

JOHN O'BRIEN

A decade ago, Senator Tom Coburn (R–OK) observed "Americans have a crazy idea, that they should get something for their money even when the money is spent by the government. It is a simple concept, and in policy-speak we call it performance-based budgeting. I know I am new in the Senate, but I am still surprised by the level of resistance in Washington to holding people accountable by measuring their performance."

Holding people accountable by measuring performance has almost become a daily routine for public administrators and it has been embedded in federal organization's business processes and policies [witness President Trump's Executive Order calling for the heads of Federal Agencies to submit reorganization plans to improve the efficiency, effectiveness and accountability of their agency]. Despite this, former Senator Coburn's observation on the resistance to holding people accountable by measuring their performance remains a viable part of our culture. Nowhere can this be seen than in the area of cyberspace where the notion of results-based performance—*cyber performance management* remains largely undefined.

Cyber Performance Management

Cyberspace is commonly defined as a domain focused on information and characterized by electronic media to store, modify and exchange information; more than traditional back-office information technology (IT) serv-

*Originally published as John O'Brien, "Thinking Strategically in the Cyber Domain," *PA Times*, October 3, 2017. Reprinted with permission of the publisher.

ice, cyberspace includes cybersecurity, cybercrime and cyber-warfare. Federal agencies have developed organizational structures around cyberspace, including the Department of State Office of the Coordinator for Cyber Issues and the Department of Homeland Security Office of Cyber, Infrastructure & Resilience Policy.

Cyber performance management could be defined as where the notion of demonstrated results through outcome-based performance measures, as conveyed in the Government Performance and Results Act of 1993 (GPRA) and the GPRA Modernization Act of 2010. While many cyberspace organizations endorse the notion of performance management in theory, the challenge is implementing a performance plan into practice that truly focuses on organizational efficiency, effectiveness, and accountability. Cyber organizations frequently fall back on more traditional performance measures based on IT services which may not address strategic-level organizational issues.

Strategic Goals and Objective in Cyber

The GPRA Modernization Act of 2010 specifies that public-sector organizations develop a strategic plan that links mission/vision statements, strategic goals, strategic objectives, performance measure and strategic initiatives that are outcome-based evidence of demonstrated results. Cyber organizations often do not do that. One could think of this in terms of a "*What-How*" model whereas agency's strategic goals and strategic objectives are the desired results [the "*What*"] while strategic initiatives and resources are the method by which results are achieved [the "*How*"].

Strategic goals are the first line of implementation towards the organization's mission and vision. They define courses of action and/or end-states that, if accomplished or achieved, will enable the organization to better support its *mission* and advance towards its *vision*. By definition, strategic goals are "strategic" in nature; that is, they are broad in scope and deal with high-level issues relevant to the organizations strategy for success (e.g., *innovation leadership, operational excellence, customer intimacy*). Strategic goals should imply that some form of *change* should be made in what the organization hope to accomplish (e.g., *improve, enhance, increase/decrease*).

Strategic objectives are an elaboration of a strategic goal, a "breakdown" of the goal.

Each strategic objective should provide greater specificity of the strategic goal that your organization is working to achieve. Strategic objectives are expressed so as to facilitate future assessment as to whether the goal was or is being achieved, are directly measurable, and are outcome/output oriented.

A common way of thinking about strategic objectives is to use the acronym "SMART":

- **S: Specific**—Objectives should specify what you want to achieve. For example, a military medical command wants to achieve 93 percent patient satisfaction in 12 months.
- **M: Measureable**—You should be able to measure whether you are meeting the objectives or not. For example, this 93 percent patient satisfaction rating over 12 months means that each month patient satisfaction can be measured against a specific target.
- **A: Achievable**—Objectives you set are really attainable. For example, is the 93 percent objective in 12 months something that can be done?
- **R: Realistic**—Can you realistically achieve the objectives given all the other constraints you have? For example, is the 93 percent objective over a 12 month period realistic given the skills and resources of the command?
- **T: Timely or Time-Bound**—The point is the objective must be clear about when you want to achieve the objective. For example, the command has set a period of 12 months to achieve the 93 percent market share target.

What Is Needed

Cyber organizations need a cyber performance management approach that includes strategic goals and objectives written with a broad, strategic-level focus on the organization. Strategic goals should be written in such a way that clearly shows a shift of responsibility for cyber to the entire organization. Strategic objectives should be measureable and contain outcome-based cyber performance metrics. Here are three examples of objectives that illustrate a well-balanced, strategic approach: (1) satisfy customer cyber requirements; (2) reduce cycle time for implementing cyber-security system upgrades; (3) increase proficiency of the cyber workforce.

55. Internet, E-mail and Computer Use Policy*

TEXAS WORKFORCE COMMISSION

Policy Statement

The use of XYZ Company (Company) electronic systems, including computers, fax machines, and all forms of Internet/intranet access, is for company business and for authorized purposes only. Brief and occasional personal use of the electronic mail system or the Internet is acceptable as long as it is not excessive or inappropriate, occurs during personal time (lunch or other breaks), and does not result in expense or harm to the Company or otherwise violate this policy.

Use is defined as "excessive" if it interferes with normal job functions, responsiveness, or the ability to perform daily job activities. Electronic communication should not be used to solicit or sell products or services that are unrelated to the Company's business; distract, intimidate, or harass coworkers or third parties; or disrupt the workplace.

Use of Company computers, networks, and Internet access is a privilege granted by management and may be revoked at any time for inappropriate conduct carried out on such systems, including, but not limited to:

- Sending chain letters or participating in any way in the creation or transmission of unsolicited commercial e-mail ("spam") that is unrelated to legitimate Company purposes;
- Engaging in private or personal business activities, including excessive use of instant messaging and chat rooms (see below);

*Public document originally published as Texas Workforce Commission, "Internet, e-mail, and Computer Use Policy," http://www.twc.state.tx.us/news/efte/internetpolicy.html.

- Accessing networks, servers, drives, folders, or files to which the employee has not been granted access or authorization from someone with the right to make such a grant;
- Making unauthorized copies of Company files or other Company data;
- Destroying, deleting, erasing, or concealing Company files or other Company data, or otherwise making such files or data unavailable or inaccessible to the Company or to other authorized users of Company systems;
- Misrepresenting oneself or the Company;
- Violating the laws and regulations of the United States or any other nation or any state, city, province, or other local jurisdiction in any way;
- Engaging in unlawful or malicious activities;
- Deliberately propagating any virus, worm, Trojan horse, trap-door program code, or other code or file designed to disrupt, disable, impair, or otherwise harm either the Company's networks or systems or those of any other individual or entity;
- Using abusive, profane, threatening, racist, sexist, or otherwise objectionable language in either public or private messages;
- Sending, receiving, or accessing pornographic materials;
- Becoming involved in partisan politics;
- Causing congestion, disruption, disablement, alteration, or impairment of Company networks or systems;
- Maintaining, organizing, or participating in non-work-related Web logs ("blogs"), Web journals, "chat rooms," or private/personal/instant messaging;
- Failing to log off any secure, controlled-access computer or other form of electronic data system to which you are assigned, if you leave such computer or system unattended;
- Using recreational games; and/or
- Defeating or attempting to defeat security restrictions on company systems and applications.

Important exception: consistent with federal law, you may use the Company's electronic systems in order to discuss with other employees the terms and conditions of your and your coworkers' employment. However, any such discussions should take place during non-duty times and should not interfere with your or your coworkers' assigned duties. You must comply with a coworker's stated request to be left out of such discussions.

Using Company electronic systems to access, create, view, transmit, or receive racist, sexist, threatening, or otherwise objectionable or illegal material, defined as any visual, textual, or auditory entity, file, or data, is strictly prohibited. Such material violates the Company anti-harassment policies and subjects the responsible employee to disciplinary action. The Company's electronic mail system, Internet access, and computer systems must not be used to harm others or to violate the laws and regulations of the United States or any other nation or any state, city, province, or other local jurisdiction in any way. Use of company resources for illegal activity can lead to disciplinary action, up to and including dismissal and criminal prosecution. The Company will comply with reasonable requests from law enforcement and regulatory agencies for logs, diaries, archives, or files on individual Internet activities, e-mail use, and/or computer use.

Unless specifically granted in this policy, any non-business use of the Company's electronic systems is expressly forbidden.

If you violate these policies, you could be subject to disciplinary action, up to and including dismissal.

Ownership and Access of Electronic Mail, Internet Access and Computer Files; No Expectation of Privacy

The Company owns the rights to all data and files in any computer, network, or other information system used in the Company and to all data and files sent or received using any company system or using the Company's access to any computer network, to the extent that such rights are not superseded by applicable laws relating to intellectual property. The Company also reserves the right to monitor electronic mail messages (including personal/private/instant messaging systems) and their content, as well as any and all use by employees of the Internet and of computer equipment used to create, view, or access e-mail and Internet content. Employees must be aware that the electronic mail messages sent and received using Company equipment or Company-provided Internet access, including web-based messaging systems used with such systems or access, are not private and are subject to viewing, downloading, inspection, release, and archiving by Company officials at all times. The Company has the right to inspect any and all files stored in private areas of the network or on individual computers or storage media in order to assure compliance with Company policies and state and federal laws. No employee may access another employee's computer, computer files, or electronic mail messages without prior authorization from either the employee or an appropriate Company official.

The Company uses software in its electronic information systems that allows monitoring by authorized personnel and that creates and stores copies of any messages, files, or other information that is entered into, received by, sent, or viewed on such systems. There is no expectation of privacy in any information or activity conducted, sent, performed, or viewed on or with Company equipment or Internet access. Accordingly, employees should assume that whatever they do, type, enter, send, receive, and view on Company electronic information systems is electronically stored and subject to inspection, monitoring, evaluation, and Company use at any time. Further, employees who use Company systems and Internet access to send or receive files or other data that would otherwise be subject to any kind of confidentiality or disclosure privilege thereby waive whatever right they may have to assert such confidentiality or privilege from disclosure. Employees who wish to maintain their right to confidentiality or a disclosure privilege must send or receive such information using some means other than Company systems or the company-provided Internet access.

The Company has licensed the use of certain commercial software application programs for business purposes. Third parties retain the ownership and distribution rights to such software. No employee may create, use, or distribute copies of such software that are not in compliance with the license agreements for the software. Violation of this policy can lead to disciplinary action, up to and including dismissal.

Confidentiality of Electronic Mail

As noted above, electronic mail is subject at all times to monitoring, and the release of specific information is subject to applicable state and federal laws and Company rules, policies, and procedures on confidentiality. Existing rules, policies, and procedures governing the sharing of confidential information also apply to the sharing of information via commercial software. Since there is the possibility that any message could be shared with or without your permission or knowledge, the best rule to follow in the use of electronic mail for non-work-related information is to decide if you would post the information on the office bulletin board with your signature.

It is a violation of Company policy for any employee, including system administrators and supervisors, to access electronic mail and computer systems files to satisfy curiosity about the affairs of others, unless such access is directly related to that employee's job duties. Employees found to have engaged in such activities will be subject to disciplinary action.

Electronic Mail Tampering

Electronic mail messages received should not be altered without the sender's permission; nor should electronic mail be altered and forwarded to another user and/or unauthorized attachments be placed on another's electronic mail message.

Policy Statement for Internet/Intranet Browser(s)

The Internet is to be used to further the Company's mission, to provide effective service of the highest quality to the Company's customers and staff, and to support other direct job-related purposes. Supervisors should work with employees to determine the appropriateness of using the Internet for professional activities and career development. The various modes of Internet/Intranet access are Company resources and are provided as business tools to employees who may use them for research, professional development, and work-related communications. Limited personal use of Internet resources is a special exception to the general prohibition against the personal use of computer equipment and software.

Employees are individually liable for any and all damages incurred as a result of violating company security policy, copyright, and licensing agreements.

All Company policies and procedures apply to employees' conduct on the Internet, especially, but not exclusively, relating to: intellectual property, confidentiality, company information dissemination, standards of conduct, misuse of company resources, anti-harassment, and information and data security.

Personal Electronic Equipment

The Company prohibits the use in the workplace of any type of camera phone, cell phone camera, digital camera, video camera, or other form of recording device to record the image or other personal information of another person, if such use would constitute a violation of a civil or criminal statute that protects the person's right to be free from harassment or from invasion of the person's right to privacy. Employees may take pictures and make recordings during non-working time in a way that does not violate such civil or criminal statutes. The Company reserves the right to report any illegal use of such devices to appropriate law enforcement authorities.

Due to the significant risk of harm to the company's electronic resources, or loss of data, from any unauthorized access that causes data loss or disruption, employees should not bring personal computers or data storage devices (such as floppy disks, CDs/DVDs, external hard drives, USB / flash drives, "smart" phones, iPods/iPads/iTouch or similar devices, laptops or other mobile computing devices, or other data storage media) to the workplace and connect them to Company electronic systems unless expressly permitted to do so by the Company. To minimize the risk of unauthorized copying of confidential company business records and proprietary information that is not available to the general public, any employee connecting a personal computing device, data storage device, or image-recording device to Company networks or information systems thereby gives permission to the Company to inspect the personal computer, data storage device, or image-recording device at any time with personnel and/or electronic resources of the Company's choosing and to analyze any files, other data, or data storage devices or media that may be within or connectable to the data-storage device in question in order to ensure that confidential company business records and proprietary information have not been taken without authorization. Employees who do not wish such inspections to be done on their personal computers, data storage devices, or imaging devices should not connect them to Company computers or networks.

Violation of this policy, or failure to permit an inspection of any device under the circumstances covered by this policy, shall result in disciplinary action, up to and possibly including immediate termination of employment, depending upon the severity and repeat nature of the offense. In addition, the employee may face both civil and criminal liability from the Company, from law enforcement officials, or from individuals whose rights are harmed by the violation.

56. Consumers Who Froze Their Credit Reports*

Julie Appleby

Some Americans who froze their credit reports following the big data breach this year at the credit-rating firm Equifax may be in for a surprise if they try to purchase insurance on the federal health law's marketplaces. That freeze could trigger a delay in the application process.

Signing up for a marketplace plan online requires consumers—especially first-time enrollees—to prove their identity by answering questions linked to their credit history. It can affect both those who are seeking a subsidy and those who are not.

But here's the rub: Consumers who have blocked access to their credit reports often "cannot get past the ID-proofing part," said Matthew Byrne, founder of the Spiralight Group, a brokerage in Dublin, Ohio.

That presents problems as the clock is ticking on open-enrollment season for the Affordable Care Act's health coverage, which ends Friday in most states.

"Your identity wasn't verified," warns the healthcare.gov page. It suggests that the applicant call "the help desk," which is run by another credit reporting firm—Experian—that works with the marketplace, or call the healthcare.gov call center, according to a screenshot sent to KHN by a broker.

That call to Experian sometimes solves the problem, if the applicant can successfully answer questions generated by the firm. Those can include queries about previous addresses, vehicles owned or current mortgage lenders, according to a 2016 report from the Center on Budget and Policy Priorities.

*Originally published as Julie Appleby, "Consumers Who Froze Their Credit Reports," *Kaiser Health News*, December 13, 2017. Reprinted with permission of the publisher.

But not always. Geoff Dellapenna said shortly after he and his broker entered some of his information onto the online application, the screen showed the warning that his identity could not be verified. They would need to make another call first.

"I just put a stop to it," said Dellapenna, 46, of Canton, Ga. "I was so frustrated with how cumbersome the whole process was being that I just could not do it."

He's now working with the broker to see if he can get small-group insurance for his company—avoiding the healthcare.gov site altogether.

There are workarounds for consumers: Applicants may have to upload or mail in documents. Some brokers advise their clients to "unfreeze" their credit reports before enrolling in the ACA by calling the three major credit rating agencies: Equifax, Experian and TransUnion. The freeze can be restored once enrolled.

But those options may take time. With the Dec. 15 deadline fast approaching, brokers say some people will be upset if they wait until the last minute.

"None of this is simple," said Kelly Fristoe, a broker in Wichita Falls, Texas.

Fristoe said the subject of credit freezes and the ID verification has come up frequently on broker forums this year. Otherwise, he added, the enrollment process has been smooth.

A spokeswoman for the Department of Health and Human Services said consumers using healthcare.gov, which serves 39 states, don't need to unfreeze their credit in order to apply, "nor do we recommend that step for identity proofing." Uploading or mailing the information will generally work, she said.

Most of the states that run their own marketplaces for ACA enrollment also use the Experian system for identity proofing, although at least two—New York and California—have modified the process to help avoid some of the delays, according to the policy report.

New York, for example, uses state motor vehicle records and other data in addition to Experian—and has a special call center that certified assisters can use to help applicants who can't pass the online verification. California also uses assisters, who are registered with the state, to check documents and upload them for applicants, the report says.

Many people heeded warnings in the early fall to monitor their credit or place a "credit freeze" on their accounts with the three major reporting firms after Equifax revealed that hackers had stolen information on 143 million Americans, including Social Security numbers.

A freeze restricts access to a consumer's credit report, which can make it more difficult for identity thieves to open new accounts using stolen information. Consumers must call or go online with each of the three credit-reporting agencies to create a freeze—and keep the PIN numbers needed to reopen the accounts.

Still, a credit freeze isn't the only reason why applicants might face an ID delay, said Spiralight's Byrne.

Sometimes applicants enter a number incorrectly—say, one digit in a Social Security number. They may have a fraud alert on their credit accounts because of a previous identity theft. Or the person is young and has little or no credit history.

But in all cases, they may get a notice saying more information is required and the application process comes to a standstill. But, so long as the process was started before the end of open enrollment, consumers should be able to finish their applications.

Once they get cleared after providing documents, applicants are eligible for a special enrollment period, said Shelby Gonzales, senior policy analyst at the Center on Budget and Policy Priorities. In that case, her advice is to telephone the "healthcare.gov call center and let them know you've been cleared for identity proofing, and they should be able to help you activate the special enrollment," she said.

Appendix A. Glossary of Cybersecurity Terms

JOAQUIN JAY GONZALEZ III

access: The ability and means to communicate with or otherwise interact with a system, to use system resources to handle information, to gain knowledge of the information the system contains, or to control system components and functions.

access control: The process of granting or denying specific requests for or attempts to: 1) obtain and use information and related information processing services; and 2) enter specific physical facilities.

access control mechanism: Security measures designed to detect and deny unauthorized access and permit authorized access to an information system or a physical facility.

active attack: An actual assault perpetrated by an intentional threat source that attempts to alter a system, its resources, its data, or its operations.

active content: Software that is able to automatically carry out or trigger actions without the explicit intervention of a user.

Advanced Persistent Threat (APT): An adversary that possesses sophisticated levels of expertise and significant resources which allow it to create opportunities to achieve its objectives by using multiple attack vectors (e.g., cyber, physical, and deception).

adversary: An individual, group, organization, or government that conducts or has the intent to conduct detrimental activities.

alert: A notification that a specific attack has been detected or directed at an organization's information systems.

antispyware software: A program that specializes in detecting and blocking or removing forms of spyware.

antivirus software: A program that monitors a computer or network to detect or identify major types of malicious code and to prevent or contain malware incidents, including removing or neutralizing the malicious code.

attack: An attempt to gain unauthorized access to system services, resources, or information, or an attempt to compromise system integrity.

attack method: The manner or technique and means an adversary may use in an assault on information or an information system.

attack path: The steps that an adversary takes or may take to plan, prepare for, and execute an attack.

attack pattern: Similar cyber events or behaviors that may indicate an attack has occurred or is occurring, resulting in a security violation or a potential security violation.

attack signature: A characteristic or distinctive pattern that can be searched for or that can be used in matching to previously identified attacks.

attack surface: The set of ways in which an adversary can enter a system and potentially cause damage.

authentication: The process of verifying the identity or other attributes of an entity (user, process, or device).

authenticity: A property achieved through cryptographic methods of being genuine and being able to be verified and trusted, resulting in confidence in the validity of a transmission, information or a message, or sender of information or a message.

authorization: A process of determining, by evaluating applicable access control information, whether a subject is allowed to have the specified types of access to a particular resource.

behavior monitoring: Observing activities of users, information systems, and processes and measuring the activities against organizational policies and rule, baselines of normal activity, thresholds, and trends.

blacklist: A list of entities that are blocked or denied privileges or access.

Blue Team: A group that defends an enterprise's information systems when mock attackers (i.e., the Red Team) attack, typically as part of an operational exercise conducted according to rules established and monitored by a neutral group (i.e., the White Team).

bot: A computer connected to the Internet that has been surreptitiously/secretly compromised with malicious logic to perform activities under remote the command and control of a remote administrator.

bot master: The controller of a botnet that, from a remote location, provides direction to the compromised computers in the botnet.

botnet: A collection of computers compromised by malicious code and controlled across a network.

bug: An unexpected and relatively small defect, fault, flaw, or imperfection in an information system or device.

ciphertext: Data or information in its encrypted form.

cloud computing: A model for enabling on-demand network access to a shared pool of configurable computing capabilities or resources (e.g., networks, servers, storage, applications, and services) that can be rapidly provisioned and released with minimal management effort or service provider interaction.

computer network defense: The actions taken to defend against unauthorized activity within computer networks.

Computer Network Defense Analysis: cybersecurity work where a person: Uses defensive measures and information collected from a variety of sources to identify, analyze, and report events that occur or might occur within the network in order to protect information, information systems, and networks from threats.

Computer Network Defense Infrastructure Support: cybersecurity work where a person: Tests, implements, deploys, maintains, reviews, and administers the infrastructure hardware and software that are required to effectively manage the computer network defense service provider network and resources; monitors network to actively remediate unauthorized activities.

Continuity of Operations Plan: A document that sets forth procedures for the continued performance of core capabilities and critical operations during any disruption or potential disruption.

critical infrastructure: The systems and assets, whether physical or virtual, so vital to society that the incapacity or destruction of such may have a debilitating impact on the security, economy, public health or safety, environment, or any combination of these matters.

cryptanalysis: The operations performed in defeating or circumventing cryptographic protection of information by applying mathematical techniques and without an initial knowledge of the key employed in providing the protection.

cryptography: The use of mathematical techniques to provide security services, such as confidentiality, data integrity, entity authentication, and data origin authentication.

Customer Service and Technical Support: cybersecurity work where a person: Addresses problems, installs, configures, troubleshoots, and provides maintenance and training in response to customer requirements or inquiries (e.g., tiered-level customer support).

cyber exercise: A planned event during which an organization simulates a cyber disruption to develop or test capabilities such as preventing, detecting, mitigating, responding to or recovering from the disruption.

cyber infrastructure: An electronic information and communications systems and services and the information contained therein.

Cyber Operations: cybersecurity work where a person: Performs activities to gather evidence on criminal or foreign intelligence entities in order to mitigate possible or real-time threats, protect against espionage or insider threats, foreign sabotage, international terrorist activities, or to support other intelligence activities.

Cyber Threat Indicator: information that is necessary to describe or identify—malicious reconnaissance, including anomalous patterns of communications that appear to be transmitted for the purpose of gathering technical information related to a cybersecurity threat or security vulnerability.

cybersecurity: The activity or process, ability or capability, or state whereby information and communications systems and the information contained therein are protected from and/or defended against damage, unauthorized use or modification, or exploitation.

cybersecurity threat: an action, not protected by the First Amendment to the Constitution of the United States, on or through an information system that may result in an unauthorized effort to adversely impact the security, availability, confidentiality, or integrity of an information system or information that is stored on, processed by, or transiting an information system.

cyberspace: The interdependent network of information technology infrastructures, that includes the Internet, telecommunications networks, computer systems, and embedded processors and controllers.

data breach: The unauthorized movement or disclosure of sensitive information to a party, usually outside the organization, that is not authorized to have or see the information.

data integrity: The property that data is complete, intact, and trusted and has not been modified or destroyed in an unauthorized or accidental manner.

data loss: The result of unintentionally or accidentally deleting data, forgetting where it is stored, or exposure to an unauthorized party.

data loss prevention: A set of procedures and mechanisms to stop sensitive data from leaving a security boundary.

data mining: The process or techniques used to analyze large sets of existing information to discover previously unrevealed patterns or correlations.

data theft: The deliberate or intentional act of stealing of information.

decipher: To convert enciphered text to plain text by means of a cryptographic system.

decode: To convert encoded text to plain text by means of a code.

decryption: The process of transforming ciphertext into its original plaintext.

Defensive Measure: an action, device, procedure, signature, technique, or other measure applied to an information system or information that is stored on, processed by, or transiting an information system that detects, prevents, or mitigates a known or suspected cybersecurity threat or security vulnerability.

Denial of Service (Dos): An attack that prevents or impairs the authorized use of information system resources or services.

digital forensics: The processes and specialized techniques for gathering, retaining, and analyzing system-related data (digital evidence) for investigative purposes.

digital signature: A value computed with a cryptographic process using a private key and then appended to a data object, thereby digitally signing the data.

disruption: An event which causes unplanned interruption in operations or functions for an unacceptable length of time.

Distributed Denial of Service (DDoS): A denial of service technique that uses numerous systems to perform the attack simultaneously.

dynamic attack surface: The automated, on-the-fly changes of an information system's characteristics to thwart actions of an adversary.

electronic signature: Any mark in electronic form associated with an electronic document, applied with the intent to sign the document.

encipher: To convert plaintext to ciphertext by means of a cryptographic system.

encode: To convert plaintext to ciphertext by means of a code.

encrypt: The generic term encompassing encipher and encode.

encryption: The process of transforming plaintext into ciphertext.

enterprise risk management: A comprehensive approach to risk management that engages people, processes, and systems across an organization to improve the quality of decision making for managing risks that may hinder an organization's ability to achieve its objectives.

exfiltration: The unauthorized transfer of information from an information system.

exploit: A technique to breach the security of a network or information system in violation of security policy.

Exploitation Analysis: cybersecurity work where a person: Analyzes collected information to identify vulnerabilities and potential for exploitation.

exposure: The condition of being unprotected, thereby allowing access to information or access to capabilities that an attacker can use to enter a system or network.

failure: The inability of a system or component to perform its required functions within specified performance requirements.

federal entity: a department or agency of the United States or any component of such department or agency.

firewall: A capability to limit network traffic between networks and/or information systems.

hacker: An unauthorized user who attempts to or gains access to an information system.

hash value: A numeric value resulting from applying a mathematical algorithm against a set of data such as a file.

hashing: process of applying a mathematical algorithm against a set of data to produce a numeric value (a 'hash value') that represents the data.

ICT supply chain threat: A man-made threat achieved through exploitation of the information and communications technology (ICT) system's supply chain, including acquisition processes.

identity and access management: The methods and processes used to manage subjects and their authentication and authorizations to access specific objects.

incident: An occurrence that actually or potentially results in adverse consequences to an information system or the information that the system processes, stores, or transmits and that may require a response action to mitigate the consequences.

incident management: The management and coordination of activities associated with an actual or potential occurrence of an event that may result in adverse consequences to information or information systems.

incident response: The activities that address the short-term, direct effects of an incident and may also support short-term recovery.

incident response plan: A set of predetermined and documented procedures to detect and respond to a cyber incident.

Information and Communications Technology: Any information technology, equipment, or interconnected system or subsystem of equipment that processes, transmits, receives, or interchanges data or information.

Information Assurance Compliance: cybersecurity work where a person: Oversees, evaluates, and supports the documentation, validation, and accreditation processes necessary to assure that new IT systems meet the organization's information assurance and security requirements.

information security policy: An aggregate of directives, regulations, rules, and practices that prescribe how an organization manages, protects, and distributes information.

information sharing: An exchange of data, information, and/or knowledge to manage risks or respond to incidents.

Information System: industrial control systems, such as supervisory control and data acquisition systems, distributed control systems, and programmable logic controllers.

Information Systems Security Operations: cybersecurity work where a person: Oversees the information assurance program of an information system in or outside the network environment; may include procurement duties (e.g., Information Systems Security Officer).

interoperability: The ability of two or more systems or components to exchange information and to use the information that has been exchanged.

intrusion: An unauthorized act of bypassing the security mechanisms of a network or information system.

intrusion detection: The process and methods for analyzing information from networks and information systems to determine if a security breach or security violation has occurred.

key: The numerical value used to control cryptographic operations, such as decryption, encryption, signature generation, or signature verification.

key pair: A public key and its corresponding private key.

keylogger: Software or hardware that tracks keystrokes and keyboard events, usually secretly, to monitor actions by the user of an information system.

macro virus: A type of malicious code that attaches itself to documents and uses the macro programming capabilities of the document's application to execute, replicate, and spread or propagate itself.

malicious applet: A small application program that is automatically downloaded and executed and that performs an unauthorized function on an information system.

malicious code: Program code intended to perform an unauthorized function or process that will have adverse impact on the confidentiality, integrity, or availability of an information system.

malicious cyber command and control: a method for unauthorized remote identification of, access to, or use of, an information system or information that is stored on, processed by, or transiting an information system.

malicious logic: Hardware, firmware, or software that is intentionally included or inserted in a system to perform an unauthorized function or process that will have adverse impact on the confidentiality, integrity, or availability of an information system.

malicious reconnaissance: a method for actively probing or passively monitoring an information system for the purpose of discerning security vulnerabilities of the information system, if such method is associated with a known or suspected cybersecurity threat.

malware: Software that compromises the operation of a system by performing an unauthorized function or process.

mitigation: The application of one or more measures to reduce the likelihood of an unwanted occurrence and/or lessen its consequences.

moving target defense: The presentation of a dynamic attack surface, increasing an adversary's work factor necessary to probe, attack, or maintain presence in a cyber target.

network resilience: The ability of a network to: (1) provide continuous operation (i.e., highly resistant to disruption and able to operate in a degraded mode if damaged); (2) recover effectively if failure does occur; and (3) scale to meet rapid or unpredictable demands.

non-federal entity: means any private entity, non–Federal government agency or department, or State, tribal, or local government (including a political subdivision, department, or component thereof).

operational exercise: action-based exercise where personnel rehearse reactions to an incident scenario, drawing on their understanding of plans and procedures, roles, and responsibilities.

passive attack: An actual assault perpetrated by an intentional threat source that attempts to learn or make use of information from a system, but does not attempt to alter the system, its resources, its data, or its operations.

penetration testing: An evaluation methodology whereby assessors search for vulnerabilities and attempt to circumvent the security features of a network and/or information system.

Personal Identifying Information: The information that permits the identity of an individual to be directly or indirectly inferred.

phishing: A digital form of social engineering to deceive individuals into providing sensitive information.

privacy: The assurance that the confidentiality of, and access to, certain information about an entity is protected.

private entity: any person or private group, organization, proprietorship, partnership, trust, cooperative, corporation, or other commercial or nonprofit entity, including an officer, employee, or agent thereof.

ransomware: This type of cyberattack infects software and locks access to data until a ransom is paid.

recovery: The activities after an incident or event to restore essential services and operations in the short and medium term and fully restore all capabilities in the longer term.

Red Team: A group authorized and organized to emulate a potential adversary's attack or exploitation capabilities against an enterprise's cybersecurity posture.

Red Team exercise: An exercise, reflecting real-world conditions, that is conducted as a simulated attempt by an adversary to attack or exploit vulnerabilities in an enterprise's information systems.

rootkit: A set of software tools with administrator-level access privileges installed on an information system and designed to hide the presence of the tools, maintain the access privileges, and conceal the activities conducted by the tools.

secret key: A cryptographic key that is used for both encryption and decryption, enabling the operation of a symmetric key cryptography scheme.

security automation: The use of information technology in place of manual processes for cyber incident response and management.

security control: the management, operational, and technical controls used to protect against an unauthorized effort to adversely affect the confidentiality, integrity, and availability of an information system or its information.

security policy: A rule or set of rules that govern the acceptable use of an organization's information and services to a level of acceptable risk and the means for protecting the organization's information assets.

security vulnerability—any attribute of hardware, software, process, or procedure that could enable or facilitate the defeat of a security control.

situational awareness: Comprehending information about the current and developing security posture and risks, based on information gathered, observation and analysis, and knowledge or experience.

Software Assurance and Security Engineering: cybersecurity work where a person: Develops and writes/codes new (or modifies existing) computer applications, software, or specialized utility programs following software assurance best practices.

spam: The abuse of electronic messaging systems to indiscriminately send unsolicited bulk messages.

spoofing: Faking the sending address of a transmission to gain illegal [unauthorized] entry into a secure system.

spyware: Software that is secretly or surreptitiously installed into an information system without the knowledge of the system user or owner.

Structured Threat Information Expression (STIX): a language for describing cyber threat information in a standard manner for the reading convenience of machines, not humans.

System Administration: cybersecurity work where a person: Installs, configures, troubleshoots, and maintains server configurations (hardware and software) to ensure their confidentiality, integrity, and availability; also manages accounts, firewalls, and patches; responsible for access control, passwords, and account creation and administration.

system integrity: The attribute of an information system when it performs its intended function in an unimpaired manner, free from deliberate or inadvertent unauthorized manipulation of the system.

Systems Security Analysis: cybersecurity work where a person: Conducts the integration/testing, operations, and maintenance of systems security.

Systems Security Architecture: work where a person: Develops system concepts and works on the capabilities phases of the systems development lifecycle; translates technology and environmental conditions (e.g., law and regulation) into system and security designs and processes.

tabletop exercise: A discussion-based exercise where personnel meet in a classroom setting or breakout groups and are presented with a scenario to validate the content of plans, procedures, policies, cooperative agreements or other information for managing an incident.

tailored trustworthy space: A cyberspace environment that provides a user with confidence in its security, using automated mechanisms to ascertain security conditions and adjust the level of security based on the user's context and in the face of an evolving range of threats.

targets: cybersecurity work where a person: Applies current knowledge of one or more regions, countries, non-state entities, and/or technologies.

threat: A circumstance or event that has or indicates the potential to exploit vulnerabilities and to adversely impact (create adverse consequences for) organizational operations, organizational assets (including information and information systems), individuals, other organizations, or society.

threat agent: An individual, group, organization, or government that conducts or has the intent to conduct detrimental activities.

threat assessment: The product or process of identifying or evaluating entities, actions, or occurrences, whether natural or man-made, that have or indicate the potential to harm life, information, operations, and/or property.

ticket: access control, data that authenticates the identity of a client or a service and, together with a temporary encryption key (a session key), forms a credential.

trojan horse: A computer program that appears to have a useful function, but also has a hidden and potentially malicious function that evades security mechanisms, sometimes by exploiting legitimate authorizations of a system entity that invokes the program.

Trusted Automated Exchange of Indicator Information (TAXII): a standard for exchanging structured cyber threat information in a trusted manner. TAXII defines services, protocols and messages to exchange cyber threat information for the detection, prevention, and mitigation of cyber threats.

virus: A computer program that can replicate itself, infect a computer without permission or knowledge of the user, and then spread or propagate to another computer.

vulnerability: A characteristic or specific weakness that renders an organization or asset (such as information or an information system) open to exploitation by a given threat or susceptible to a given hazard.

Vulnerability Assessment and Management: Conducts assessments of threats and vulnerabilities, determines deviations from acceptable configurations, enterprise or local policy, assesses the level of risk, and develops and/or recommends appropriate mitigation countermeasures in operational and non-operational situations.

weakness: A shortcoming or imperfection in software code, design, architecture, or deployment that, under proper conditions, could become a vulnerability or contribute to the introduction of vulnerabilities.

White Team: A group responsible for refereeing an engagement between a Red Team of mock attackers and a Blue Team of actual defenders of information systems.

whitelist: A list of entities that are considered trustworthy and are granted access or privileges.

worm: A self-replicating, self-propagating, self-contained program that uses networking mechanisms to spread itself.

References

niccs.us-cert.gov/glossary
U.S. Department of Commerce
U.S. Department of Homeland Security

Appendix B. Presidential Executive Order on Strengthening the Cybersecurity of Federal Networks and Critical Infrastructure

https://www.whitehouse.gov/presidential-actions/presidential-executive-order-strengthening-cybersecurity-federal-networks-critical-infrastructure/

By the authority vested in me as President by the Constitution and the laws of the United States of America, and to protect American innovation and values, it is hereby ordered as follows:

Section 1. Cybersecurity of Federal Networks.

(a) Policy. The executive branch operates its information technology (IT) on behalf of the American people. Its IT and data should be secured responsibly using all United States Government capabilities. The President will hold heads of executive departments and agencies (agency heads) accountable for managing cybersecurity risk to their enterprises. In addition, because risk management decisions made by agency heads can affect the risk to the executive branch as a whole, and to national security, it is also the policy of the United States to manage cybersecurity risk as an executive branch enterprise.

(b) Findings.

(i) Cybersecurity risk management comprises the full range of activities undertaken to protect IT and data from unauthorized access and other cyber threats, to maintain awareness of cyber threats, to detect anomalies and incidents adversely affecting IT and data, and to mitigate the impact of, respond to, and recover from incidents. Information sharing facilitates and supports all of these activities.

(ii) The executive branch has for too long accepted antiquated and difficult-to-defend IT.

(iii) Effective risk management involves more than just protecting IT and data currently in place. It also requires planning so that maintenance, improvements, and modernization occur in a coordinated way and with appropriate regularity.

(iv) Known but unmitigated vulnerabilities are among the highest cybersecurity risks faced by executive departments and agencies (agencies). Known vulnerabilities include using operating systems or hardware beyond the vendor's support lifecycle, declining to implement a vendor's security patch, or failing to execute security-specific configuration guidance.

(v) Effective risk management requires agency heads to lead integrated teams of senior executives with expertise in IT, security, budgeting, acquisition, law, privacy, and human resources.

(c) Risk Management.

(i) Agency heads will be held accountable by the President for implementing risk management measures commensurate with the risk and magnitude of the harm that would result from unauthorized access, use, disclosure, disruption, modification, or destruction of IT and data. They will also be held accountable by the President for ensuring that cybersecurity risk management processes are aligned with strategic, operational, and budgetary planning processes, in accordance with chapter 35, subchapter II of title 44, United States Code.

(ii) Effective immediately, each agency head shall use The Framework for Improving Critical Infrastructure Cybersecurity (the Framework) developed by the National Institute of Standards and Technology, or any successor document, to manage the agency's cybersecurity risk. Each agency head shall provide a risk management report to the Secretary of Homeland Security and the Director of the Office of Management and Budget (OMB) within 90 days of the date of this order. The risk management report shall:

(A) document the risk mitigation and acceptance choices made by each agency head as of the date of this order, including:

(1) the strategic, operational, and budgetary considerations that informed those choices; and

(2) any accepted risk, including from unmitigated vulnerabilities; and

(B) describe the agency's action plan to implement the Framework.

(iii) The Secretary of Homeland Security and the Director of OMB, consistent with chapter 35, subchapter II of title 44, United States Code, shall jointly assess each agency's risk management report to determine whether the risk mitigation and acceptance choices set forth in the reports are appropriate and sufficient to manage the cybersecurity risk to the executive branch enterprise in the aggregate (the determination).

(iv) The Director of OMB, in coordination with the Secretary of Homeland Security, with appropriate support from the Secretary of Commerce and the Administrator of General Services, and within 60 days of receipt of the agency risk management reports outlined in subsection (c)(ii) of this section, shall submit to the President, through the Assistant to the President for Homeland Security and Counterterrorism, the following:

(A) the determination; and

(B) a plan to:

(1) adequately protect the executive branch enterprise, should the determination identify insufficiencies;

(2) address immediate unmet budgetary needs necessary to manage risk to the executive branch enterprise;

(3) establish a regular process for reassessing and, if appropriate, reissuing the determination, and addressing future, recurring unmet budgetary needs necessary to manage risk to the executive branch enterprise;

(4) clarify, reconcile, and reissue, as necessary and to the extent permitted by law, all policies, standards, and guidelines issued by any agency in furtherance of chapter 35, subchapter II of title 44, United States Code, and, as necessary and to the extent permitted by law, issue policies, standards, and guidelines in furtherance of this order; and

(5) align these policies, standards, and guidelines with the Framework.

(v) The agency risk management reports described in subsection (c)(ii) of this section and the determination and plan described in subsections (c)(iii) and (iv) of this section may be classified in full or in part, as appropriate.

(vi) Effective immediately, it is the policy of the executive branch to build and maintain a modern, secure, and more resilient executive branch IT architecture.

(A) Agency heads shall show preference in their procurement for shared IT services, to the extent permitted by law, including email, cloud, and cybersecurity services.

(B) The Director of the American Technology Council shall coordinate a report to the President from the Secretary of Homeland Security, the Director of OMB, and the Administrator of General Services, in consultation with the Secretary of Commerce, as appropriate, regarding modernization of Federal IT. The report shall:

(1) be completed within 90 days of the date of this order; and

(2) describe the legal, policy, and budgetary considerations relevant to—as well as the technical feasibility and cost effectiveness, including timelines and milestones, of—transitioning all agencies, or a subset of agencies, to:

(aa) one or more consolidated network architectures; and

(bb) shared IT services, including email, cloud, and cybersecurity services.

(C) The report described in subsection (c)(vi)(B) of this section shall assess the effects of transitioning all agencies, or a subset of agencies, to shared IT services with respect to cybersecurity, including by making recommendations to ensure consistency with section 227 of the Homeland Security Act (6 U.S.C. 148) and compliance with policies and practices issued in accordance with section 3553 of title 44, United States Code. All agency heads shall supply such information concerning their current IT architectures and plans as is necessary to complete this report on time.

(vii) For any National Security System, as defined in section 3552(b)(6) of title 44, United States Code, the Secretary of Defense and the Director of National Intelligence, rather than the Secretary of Homeland Security and the Director of OMB, shall implement this order to the maximum extent feasible and appropriate. The Secretary of Defense and the Director of National Intelligence shall provide a report to the Assistant to the President for National Security Affairs and the Assistant to the President for Homeland Security and Counterterrorism describing their implementation of subsection (c) of this section within 150 days of the date of this order. The report described in this subsection shall include a justification for any deviation from the requirements of subsection (c), and may be classified in full or in part, as appropriate.

Sec. 2. Cybersecurity of Critical Infrastructure.

(a) Policy. It is the policy of the executive branch to use its authorities and capabilities to support the cybersecurity risk management efforts of the owners and operators of the Nation's critical infrastructure (as defined in section 5195c(e) of title 42, United States Code) (critical infrastructure entities), as appropriate.

(b) Support to Critical Infrastructure at Greatest Risk. The Secretary of Homeland Security, in coordination with the Secretary of Defense, the Attorney General, the Director of National Intelligence, the Director of the Federal Bureau of Investigation, the heads of appropriate sector-specific agencies, as defined in Presidential Policy Directive 21 of February 12, 2013 (Critical Infrastructure Security and Resilience) (sector-specific agencies), and all other appropriate agency heads, as identified by the Secretary of Homeland Security, shall:

(i) identify authorities and capabilities that agencies could employ to support the cybersecurity efforts of critical infrastructure entities identified pursuant to section 9 of Executive Order 13636 of February 12, 2013 (Improving Critical Infrastructure Cybersecurity), to be at greatest risk of attacks that could reasonably result in catastrophic regional or national effects on public health or safety, economic security, or national security (section 9 entities);

(ii) engage section 9 entities and solicit input as appropriate to evaluate whether and how the authorities and capabilities identified pursuant to subsection (b)(i) of this section might be employed to support cybersecurity risk management efforts and any obstacles to doing so;

(iii) provide a report to the President, which may be classified in full or in part, as appropriate, through the Assistant to the President for Homeland Security and Counterterrorism, within 180 days of the date of this order, that includes the following:

(A) the authorities and capabilities identified pursuant to subsection (b)(i) of this section;

(B) the results of the engagement and determination required pursuant to subsection (b)(ii) of this section; and

(C) findings and recommendations for better supporting the cybersecurity risk management efforts of section 9 entities; and

(iv) provide an updated report to the President on an annual basis thereafter.

(c) Supporting Transparency in the Marketplace. The Secretary of Homeland Security, in coordination with the Secretary of Commerce, shall provide a report to the President, through the Assistant to the President for Homeland Security and Counterterrorism, that examines the sufficiency of existing Federal policies and practices to promote appropriate market transparency of cybersecurity risk management practices by critical infrastructure entities, with a focus on publicly traded critical infrastructure entities, within 90 days of the date of this order.

(d) Resilience Against Botnets and Other Automated, Distributed Threats. The Secretary of Commerce and the Secretary of Homeland Security shall jointly lead an open and transparent process to identify and promote action by appropriate stakeholders to improve the resilience of the internet and communications ecosystem and to encourage collaboration with the goal of dramatically reducing threats perpetrated by automated and distributed attacks (e.g., botnets). The Secretary of Commerce and the Secretary of Homeland Security shall consult with the Secretary of Defense, the Attorney General, the Director of the Federal Bureau of Investigation, the heads of sector-specific agencies, the Chairs of the Federal Communications Commission and Federal Trade Commission, other interested agency heads, and appropriate stakeholders in carrying out this subsection. Within 240 days of the date of this order, the Secretary of Commerce and the Secretary of Homeland Security shall make publicly available a preliminary report on this effort. Within 1 year of the date of this order, the Secretaries shall submit a final version of this report to the President.

(e) Assessment of Electricity Disruption Incident Response Capabilities. The Secretary of Energy and the Secretary of Homeland Security, in consultation with the Director of National Intelligence, with State, local, tribal, and territorial governments, and with others as appropriate, shall jointly assess:

(i) the potential scope and duration of a prolonged power outage associated with a significant cyber incident, as defined in Presidential Policy Directive 41 of July 26, 2016 (United States Cyber Incident Coordination), against the United States electric subsector;

(ii) the readiness of the United States to manage the consequences of such an incident; and

(iii) any gaps or shortcomings in assets or capabilities required to mitigate the consequences of such an incident.

The assessment shall be provided to the President, through the Assistant to the President for Homeland Security and Counterterrorism, within 90 days of the date of this order, and may be classified in full or in part, as appropriate.

(f) Department of Defense Warfighting Capabilities and Industrial Base. Within 90 days of the date of this order, the Secretary of Defense, the Secretary of Homeland Security, and the Director of the Federal Bureau of Investigation, in coordination with the Director of National Intelligence, shall provide a report to the President, through the Assistant to the President for National Security Affairs and the Assistant to the President for Homeland Security and Counterterrorism, on cybersecurity risks facing the defense industrial base, including its supply chain, and United States military platforms, systems, networks, and capabilities, and recommendations for mitigating these risks. The report may be classified in full or in part, as appropriate.

Sec. 3. Cybersecurity for the Nation.

(a) Policy. To ensure that the internet remains valuable for future generations, it is the policy of the executive branch to promote an open, interoperable, reliable, and secure internet that fosters efficiency, innovation, communication, and economic prosperity, while respecting privacy and guarding against disruption, fraud, and theft. Further, the United States seeks to support the growth and sustainment of a workforce that is skilled in cybersecurity and related fields as the foundation for achieving our objectives in cyberspace.

(b) Deterrence and Protection. Within 90 days of the date of this order, the Secretary of State, the Secretary of the Treasury, the Secretary of Defense, the Attorney General, the Secretary of Commerce, the Secretary of Homeland Security, and the United States Trade Representative, in coordination with the Director of National Intelligence, shall jointly submit a report to the President, through the Assistant to the President for National Security Affairs and the Assistant to the President for Homeland Security and Counterterrorism, on the Nation's strategic options for deterring adversaries and better protecting the American people from cyber threats.

(c) International Cooperation. As a highly connected nation, the United States is especially dependent on a globally secure and resilient internet and must work with allies and other partners toward maintaining the policy set forth in this section. Within 45 days of the date of this order, the Secretary of State, the Secretary of the Treasury, the Secretary of Defense, the Secretary of Commerce, and the Secretary of Homeland Security, in coordination with the Attorney General and the Director of the Federal Bureau of Investigation, shall submit reports to the President on their international cybersecurity priorities, including those concerning investigation, attribution, cyber threat information sharing, response, capacity building, and cooperation. Within 90 days of the submission of the reports, and in coordination with the agency heads listed in this subsection, and any other agency heads as appropriate, the Secretary of State shall provide a report to the President, through the Assistant to the President for Homeland Security and Counterterrorism, documenting an engagement strategy for international cooperation in cybersecurity.

(d) Workforce Development. In order to ensure that the United States maintains a long-term cybersecurity advantage:

(i) The Secretary of Commerce and the Secretary of Homeland Security, in consultation with the Secretary of Defense, the Secretary of Labor, the Secretary of Education, the Director of the Office of Personnel Management, and other agencies identified jointly by the Secretary of Commerce and the Secretary of Homeland Security, shall:

(A) jointly assess the scope and sufficiency of efforts to educate and train the American cybersecurity workforce of the future, including cybersecurity-related education curricula, training, and apprenticeship programs, from primary through higher education; and

(B) within 120 days of the date of this order, provide a report to the President, through the Assistant to the President for Homeland Security and Counterterrorism, with findings and recommendations regarding how to support the growth and sustainment of the Nation's cybersecurity workforce in both the public and private sectors.

(ii) The Director of National Intelligence, in consultation with the heads of other agencies identified by the Director of National Intelligence, shall:

(A) review the workforce development efforts of potential foreign cyber peers in order to help identify foreign workforce development practices likely to affect long-term United States cybersecurity competitiveness; and

(B) within 60 days of the date of this order, provide a report to the President through the Assistant to the President for Homeland Security and Counterterrorism on the findings of the review carried out pursuant to subsection (d)(ii)(A) of this section.

(iii) The Secretary of Defense, in coordination with the Secretary of Commerce, the Secretary of Homeland Security, and the Director of National Intelligence, shall:

(A) assess the scope and sufficiency of United States efforts to ensure that the United States maintains or increases its advantage in national-security-related cyber capabilities; and

(B) within 150 days of the date of this order, provide a report to the President, through the Assistant to the President for Homeland Security and Counterterrorism, with findings and recommendations on the assessment carried out pursuant to subsection (d)(iii)(A) of this section.

(iv) The reports described in this subsection may be classified in full or in part, as appropriate.

Sec. 4. Definitions. For the purposes of this order:

(a) The term "appropriate stakeholders" means any non-executive-branch person or entity that elects to participate in an open and transparent process established by the Secretary of Commerce and the Secretary of Homeland Security under section 2(d) of this order.

(b) The term "information technology" (IT) has the meaning given to that term in section 11101(6) of title 40, United States Code, and further includes hardware and software systems of agencies that monitor and control physical equipment and processes.

(c) The term "IT architecture" refers to the integration and implementation of IT within an agency.

(d) The term "network architecture" refers to the elements of IT architecture that enable or facilitate communications between two or more IT assets.

Sec. 5. General Provisions. (a) Nothing in this order shall be construed to impair or otherwise affect:

(i) the authority granted by law to an executive department or agency, or the head thereof; or

(ii) the functions of the Director of OMB relating to budgetary, administrative, or legislative proposals.

(b) This order shall be implemented consistent with applicable law and subject to the availability of appropriations.

(c) All actions taken pursuant to this order shall be consistent with requirements and authorities to protect intelligence and law enforcement sources and methods. Nothing in this order shall be construed to supersede measures established under authority of law to protect the security and integrity of specific activities and associations that are in direct support of intelligence or law enforcement operations.

(d) This order is not intended to, and does not, create any right or benefit, substantive or procedural, enforceable at law or in equity by any party against the United States, its departments, agencies, or entities, its officers, employees, or agents, or any other person.

DONALD J. TRUMP

THE WHITE HOUSE,
May 11, 2017

Appendix C. City and County of San Francisco Cybersecurity Policy

The City and County of San Francisco is dedicated towards building a strong cybersecurity program to support, maintain, and secure critical infrastructure and data systems. The following policy is intended to establish key elements to a citywide cybersecurity program.

Purpose and Scope

The COIT Cybersecurity Policy lays the foundation for the City's Cybersecurity Program as a whole and demonstrates executive level support for the program. Cybersecurity operations across the City are in different stages of operation. The Cybersecurity Policy will help build the City's Cybersecurity Program in order to:

- protect our connected critical infrastructure
- protect the sensitive information placed in our trust
- manage risk
- continuously improve our ability to detect cybersecurity events
- contain and eradicate compromises, restoring information resources to a secured and operational status
- ensure risk treatment is adequate and in alignment with the criticality of the information resource
- facilitate awareness of risk to our operations within the context of cybersecurity

The requirements identified in this policy apply to all information resources operated by or for the City and County of San Francisco and its component departments and commissions. Elected officials, employees, consultants, and vendors working on behalf of the City and County of San Francisco are required to comply with this policy.

Policy Statement

The Cybersecurity Policy lays the foundation for the City and County of San Francisco's Cybersecurity Program as a whole and demonstrates executive level support for the program.

The COIT Cybersecurity Policy requires all departments to:

1. Adopt a cybersecurity framework as a basis to build their cybersecurity program. The City recommends adopting the National Institute of Standards and Technology (NIST) Cybersecurity Framework as a methodology to secure information resources.
2. Provide management level support to conduct cybersecurity operations.
3. Appoint Cybersecurity Officers or a security liaisons to coordinate cybersecurity efforts.
4. Participate in citywide cybersecurity round table meetings.

These requirements will support the operations of the City's cybersecurity operations.

Framework Requirements

The Cybersecurity Policy requires all departments to adopt a cybersecurity framework to guide their operations.

In order to adequately protect information resources, systems and data must be properly categorized based on information sensitivity and criticality to operations. A risk based methodology standardizes the security architecture, creates a common understanding of shared or transferred risk when systems and infrastructure are interconnected, and makes securing systems and data more straightforward.

The NIST framework provides five elements to a cybersecurity program as follows:

- Identify: Develop the organizational understanding to manage cyber security risk to systems, assets, data, and capabilities.
- Protect: Develop and implement the appropriate safeguards to ensure delivery of infrastructure services.
- Detect: Develop and implement the appropriate activities to identify the occurrence of a cyber security event.
- Respond: Develop and implement the appropriate activities to respond to a cyber security event.

- Recover: Develop and implement the appropriate activities to maintain plans for resilience and to restore any capabilities or services that were impaired by a cyber security event.

Departments in consultation with the City Chief Information Security Officer (CISO) may choose alternatives to the NIST Cybersecurity Framework but all departments shall implement or consume central standards and services from their respective framework, such as access control and management, risk assessment and management, awareness and training, and data classification.

To ensure their cybersecurity programs comply with the Cybersecurity Framework and the risk based approach, the City Services Auditor will conduct readiness assessments to measure implementation.

Readiness assessments will align with the NIST framework and enable departments to determine their current cyber security capabilities, set individual goals for a target state, and establish a plan for improving and maintaining cyber security programs. Readiness assessments will also assist the Department of Technology and the Controller in the efficient and effective planning of cyber security activities.

Roles and Responsibilities

1. Department Heads shall:

a. Promote a culture of cybersecurity awareness and compliance to City's cybersecurity policy. Department heads must remind employees and contractors in their departments about the City's Cybersecurity policies, standards, procedures, guidelines and best practices.

b. To the extent possible, attempt to budget and staff the cybersecurity function for systems procured, operated, or contracted by their respective departments to ensure that all systems and the data contained by them are protected in accordance with the category / classification of the data and systems.

c. Designate a cybersecurity officer or liaisons in the case of smaller departments. Departments should consult with the City's Chief Information Security Officer (CISO) to determine if their information technology activities warrant the appointment of a Department Cybersecurity Officer or if a Security Liaison is adequate.

d. When appropriate, consult with the City CISO office when gathering the requirements for new information systems to ensure that the security design is vetted before selection and deployment.

2. Department Cybersecurity Security Officers / Liaisons shall:

a. Ensure information resources are properly protected through risk treatment strategies that meet the acceptable risk threshold for the category/classification of the information resource.

b. Inform the City CISO when there is an event which compromises the confidentiality, integrity, or availability of a system or data involving Personally Identifiable Information, Regulatory Protected Information (such as HIPAA or Social Security Numbers), and/or data that is not considered public as soon as practical.

c. Participate in the citywide cybersecurity round table meetings.

3. City Chief Information Security Officer (CISO) shall:

a. Ensure that Department, Commission, and the Centralized Information Technology Cybersecurity Programs employ a risk based assessment and treatment program, and regularly report the status of CCSF's residual risk profile to City leadership.

b. Develop and maintain a centralized incident response program capable of addressing major compromises of CCSF information resources.

c. Establish and maintain a Security Operations Center with the capability to identify, protect, detect, respond, and recover from attacks against CCSF information resources.

d. Organize citywide round table cybersecurity meetings.

4. COIT and Mayor's Budget Office shall:

a. Adequately support cybersecurity operations.

5. Chief Data Officer shall:

a. Work with the City CISO to develop and maintain an information classification system and support departments in their data classification efforts.

6. City Services Auditor shall:

a. Support City cybersecurity efforts with regular readiness assessments and assist in the development and exercise of cybersecurity audit controls.

7. CCSF Employees, contractors, and vendors shall:

a. Comply with cybersecurity practices and Acceptable Use Agreement, and timely report any incidents to the appropriate officials.

Compliance

To the extent resources allow:

1. Department Heads are responsible for ensuring that systems procured, operated, or contracted by their respective department or commission meet the appropriate security protections required by the system's risk category /classification, in addition to any regulatory requirements.
2. Employees, consultants, and vendors shall ensure that information resources are appropriately and securely utilized, administered, and operated while authorized access is granted, according to the Acceptable Use Policy.

Exceptions

No exceptions will be approved to this policy.

Authorization

SEC. 22A.3. Of the City's Administrative Code states, "COIT shall review and approve the recommendations of the City CIO for ICT standards, policies and procedures to enable successful development, operation, maintenance, and support of the City's ICT."

References

NIST Cybersecurity Framework Website—http://www.nist.gov/cyberframework/

Definitions

For a list of definitions please refer to:
http://nvlpubs.nist.gov/nistpubs/ir/2013/NIST.IR.7298r2.pdf

Appendix D. LaPorte County, Indiana, IT Computer Security Policy

http://www.laportecounty.org/InformationTechnology/ComputerSecurity/ (April 23, 2018)

Objective: Almost all LaPorte County Government business and administrative functions involve the use of computer or telecommunication technologies. Information is processed and stored in vast amounts on minicomputer and microcomputer systems. It is the responsibility of every LaPorte County Government employee and contract worker to safeguard the information and the physical assets of these systems. Computer security procedures are intended to reduce or eliminate threats to computer systems and electronic information. Many of these threats do not result from malicious intent; rather they stem from basic human error. Care and awareness are the two most significant safeguards. All employees and contract staff must know what is and is not allowed in the access to and the use of computer systems and equipment.

1.1 General Security Guidelines: LaPorte County Government will develop and maintain policies and controls to ensure the security of computing and telecommunication equipment, the physical premises housing the equipment and the data used, stored or produced on the equipment. These policies and controls will be approved by the LaPorte County Data Board and LaPorte County Commissioners. The Information Technology Department will develop and maintain these policies on behalf of the government. LaPorte County Government Departments may develop supplemental policies and controls to accommodate specific requirements. These policies may not compromise government policies and controls. Roles and Responsibilities: LaPorte County Information Technology is responsible for implementing and enforcing adequate computer security policies throughout the organization. LaPorte County Information Technology is responsible for ensuring that an adequate level of security and backup exists for all data whether processed or stored in-house or externally. LaPorte County Information Technology is responsible for ensuring that all of its automated processes are designed, developed and tested so that they function accurately and effectively. LaPorte County Information Technology is responsible for ensuring that all personnel, whether employed by LaPorte County Government or under contract to a department, are made aware of the appropriate security policies and procedures and of their responsibility to conform to those policies and procedures. LaPorte County Information Technology is responsible for ensuring that all computing facilities processing LaPorte County Government information comply with LaPorte County Government security specifications. LaPorte County Information Technology is responsible for ensuring that all staff receive adequate training in the use of hardware and software required for the performance of their jobs. LaPorte County Information Technology is responsible for ensuring that all software installed on LaPorte County Government computers is properly licensed and authorized. Each Department Head is delegated responsibility and authority to implement and enforce these policies within their own department, following LaPorte County's DISCIPLINARY WARNING PROCEDURE in the PERSONNEL POLICY MANUAL, wherever and whenever it is in their control to do so.

1.2 Security Awareness Policy: The Information Technology Department and LaPorte County Government Department Heads are responsible for communicating computer security policies and procedures and for promoting and monitoring their use. Standards and Procedures: The Information Technology Department will develop and maintain the Computer Security Policies and Procedures. The Information Technology Department will review the Computer Security Policies and Procedures on an annual basis. A copy of the Computer Security Policies and Procedures will be provided to each LaPorte County Government Department by the Information Technology Department. Department Heads will ensure that all employees and contract workers in their departments are aware of and have access to the Computer Security Policies and Procedures. The Information Technology Systems Department working with Internal Audit, will report periodically to department heads on the level of adherence to the Computer Security Policies and Procedures.

1.3 Physical Security Policy: Information Technology will develop and observe standards and procedures to ensure security of the physical premises and computing equipment. Limitations: Security for equipment such as personal computers, printers, modems, etc., which is maintained outside the physical control of Information Technology is the responsibility of the

LaPorte County Government Department where that equipment resides. Standards and Procedures:Access to computer and server rooms will be limited to staff who require access for the normal performance of their job. Offices where Equipment is housed must be locked during non business hours. Equipment housed in open areas should be attached to an immovable object by a security cable if possible. Electrical power protection devices to suppress surges, reduce static, and provide battery backup in the event of a power failure should be used as necessary. Equipment which is to be removed from LaPorte County Government property daily or on occasion must have prior approval from the Department Head and Information Technology.

1.4 Network Security Policy: Information Technology will develop and observe standards and procedures to maintain security on all of its computer networks to protect the security of LaPorte County Government data and of access to LaPorte County Government computer systems. Standards and Procedures: Information Technology will ensure that the software security implemented on the networks it manages is installed and functioning correctly. Information Technology will monitor network security on a regular basis. Adequate information concerning network traffic and activity will be logged to ensure that breaches in network security can be detected. Information Technology will implement and maintain procedures to provide adequate protection from intrusion into LaPorte County Government's computer systems from external sources. Any computer containing sensitive data will be secured from unauthorized access by network-level security procedures. No computer that is connected to the network can have stored, on its disk(s) or in its memory, information that would permit access to other parts of the network. For example, scripts used in accessing a remote host may not contain passwords.

1.5 Data Security Policy: Information Technology will develop and observe standards and procedures to ensure the security of technical and user data. Limitations:Security for the data stored on computer systems must be determined by the owner of the data. The standards and procedures herein should be adhered to accordingly. Passwords: Each user should be assigned and be responsible for their own unique user ID and password. A password should be known only to the authorized user of that ID and password. Passwords should be treated as confidential information. They should not be written down or shared with other users. Users should select passwords that conform to standards, as to size and characters used, and cannot be easily guessed by other users. User access should also be restricted to only those functions they are authorized to perform. Confidential data should be protected by passwords which are known only to authorized personnel. Passwords will be changed periodically to maintain security. Department Heads must notify Information Technology of personnel leaving their department who have terminated employment or have been assigned to other duties. Information Technology will then delete that user ID or adjust access rights as directed by the new Department Head. Standards and Procedures: Data encryption techniques should be used when highly confidential information is stored. If a Department Head feels encryption is necessary, contact Information Technology Systems for assistance.The required data security level, as determined by the owner, must be retained when the data is moved or copied to another system. Confidential documents must have in the header or foot of the document words stating that the document is confidential. Example: LAPORTE COUNTY CIRCUIT COURT CONFIDENTIAL. Printed reports containing confidential data must be stored and discarded appropriately. Determining confidentiality is determined by owning department. Users must sign off their terminal when leaving it unattended.

1.6 Personal Computer Security Policy: Information Technology will develop and observe standards and procedures to ensure adequate security for personal computers and the applications and data stored on them.

About the Contributors

Affiliations are as of the times the works were written.

Lalit **Ahluwalia** is Accenture's health and public-sector security lead for North America.

Julie **Appleby** is a senior correspondent with *Kaiser Health News.*

Marcus **Banks** is a journalist with prior experience as an academic library administrator.

Mary Ann **Barton** is a senior staff writer for *NACo County News.*

Reggie **Best** is the chief product officer with Lumeta.

Chelsea **Binns** is an assistant professor in the department of criminal justice, legal studies and homeland security at St. John's University in New York.

Harry **Black** is a city manager of Cincinnati, Ohio.

Kelsey **Brewer** is a policy manager at the Association of California Cities-Orange County.

Sharon L. **Cardash** is the associate director, Center for Cyber and Homeland Security, George Washington University.

Fred H. **Cate** is the Distinguished Professor and C. Ben Dutton Professor of Law, Indiana University.

Subrata **Chakrabarti** is the vice president of product marketing and strategy at Anaplan.

Frank J. **Cilluffo** is the Associate Vice President and Director, Center for Cyber and Homeland Security, George Washington University.

Gerald **Cliff** is the research director, National White Collar Crime Center, Fairmont, West Virginia.

Ariel **Cohen** is a *NACo County News* special correspondent.

Caitlin **Cowart** is the community and public relations manager for San Antonio Public Library.

Benjamin **Dean** is the Fellow for Internet Governance and Cyber-Security, School of International and Public Affairs, Columbia University.

Dorothy **Denning** is an Emeritus Distinguished Professor of Defense Analysis, Naval Postgraduate School.

Federal Bureau of Investigation is an intelligence-driven and threat-focused national security organization with both intelligence and law enforcement responsibilities.

Cory **Fleming** is a senior technical director and program specialist at the International City/County Management Association.

Anne **Ford** is editor-at-large for *American Libraries*.

Richard **Forno** is a senior lecturer, Cybersecurity & Internet Researcher, University of Maryland, Baltimore County.

Joaquin Jay **Gonzalez** III is Mayor George Christopher Professor of Public Administration at the Edward S. Ageno School of Business of Golden Gate University.

Elana **Gordon** is a staff writer for *Kaiser Health News*.

Government Finance Officers Association is a professional association of approximately 19,000 state, provincial, and local government finance officers in the United States and Canada.

James H. **Hamlyn-Harris** is a senior lecturer, Computer Science and Software Engineering, Swinburne University of Technology.

William **Hatcher** is an associate professor and director of the master of public administration program at Augusta University.

Gus "Ira" **Hunt** is the managing director and cybersecurity practice lead for Accenture Federal Services.

International City/County Management Association is an association representing professionals in local government management based in Washington, D.C.

Intersector Project is a non-profit organization that empowers practitioners in the business, government, and non-profit sectors to collaborate to solve problems that cannot be solved by one sector alone.

Kaiser Health News is an editorial independent health news service and part of the Kaiser Family Foundation.

Larry **Karisny** is the director of ProjectSafety.org, an adviser, consultant, speaker and writer supporting advanced cybersecurity technologies in both the public and private sectors.

Roger L. **Kemp** is the Distinguished Adjunct Professor of Public Administration at the Department of Public Administration of Golden Gate University.

Nir **Kshetri** is a professor of management at the University of North Carolina at Greensboro.

Theodore J. **Kury** is the Director of Energy Studies, University of Florida.

Daniel J. **Lohrmann** is an internationally recognized cybersecurity leader, technologist, keynote speaker and author.

Brian **McLaughlin** is an adjunct faculty member with the Department of Public Administration at Villanova University.

Microsoft is a U.S. corporation that develops, manufactures, licenses, supports and sells computer software, consumer electronics, personal computers, and services.

Susan **Miller** is the executive editor at GCN.

Shaun **Mulholland** is the town administrator, Allenstown, New Hampshire.

Mary Scott **Nabers** is the president and CEO of Strategic Partnerships Inc., a business development company specializing in government contracting and procurement consulting throughout the U.S.

John **O'Brien** is an associate professor in the Information Strategies Department of the College of Information and Cyberspace (CIS) at the National Defense University.

Office of the Director of National Intelligence is a cabinet-level organization tasked to lead the Intelligence Community (IC) in intelligence integration.

Martha **Perego** is the Director of Ethics and Team Leader for Membership and Professional Development at the International City/County Management Association.

Elaine S. **Povich** covers consumer affairs for *Stateline* and has done freelance work for *The Fiscal Times*, *Governing* and other publications.

Scott **Shackelford** is an associate professor of business law and ethics, director of the Ostrom Workshop Program on Cybersecurity and Internet Governance at Indiana University.

Alan **Shark** is the executive director and CEO of Public Technology Institute (PTI).

Margaret **Steen** is a contributor to *Government Technology*.

Jeremy **Straub** is an assistant professor of Computer Science, North Dakota State University.

Texas Workforce Commission is a state-level governmental agency that provides unemployment benefits and services related to employment to eligible individuals and businesses.

U.S. Computer Emergency Readiness Team is an organization within the U.S. Department of Homeland Security's National Protection and Programs Directorate.

U.S. Department of Defense is to provide a lethal Joint Force to defend the security of the United States and sustain American influence abroad.

U.S. Department of Homeland Security is a cabinet-level department of the United States federal government with responsibilities in public security.

U.S. Department of Justice is the federal department responsible for the enforcement of the law and administration of justice in the United States.

Arun **Vishwanath** is an associate professor of communication, University at Buffalo, The State University of New York.

Colin **Wood** writes for statesscoop.com and was previously with *Government Technology* and *Emergency Management*.

Index

www.ingramcontent.com/pod-product-compliance
Lightning Source LLC
LaVergne TN
LVHW042123070326
832902LV00036B/571
* 9 7 8 1 4 7 6 6 7 4 4 0 7 *